Arman Mohseni

Development of Endoscopic Stereoscopic PIV for Investigating the IGV-Impeller Interaction in a Centrifugal Compressor

Arman Mohseni

Development of Endoscopic Stereoscopic PIV for Investigating the IGV-Impeller Interaction in a Centrifugal Compressor

Revised Edition

Südwestdeutscher Verlag für Hochschulschriften

Imprint

Any brand names and product names mentioned in this book are subject to trademark, brand or patent protection and are trademarks or registered trademarks of their respective holders. The use of brand names, product names, common names, trade names, product descriptions etc. even without a particular marking in this work is in no way to be construed to mean that such names may be regarded as unrestricted in respect of trademark and brand protection legislation and could thus be used by anyone.

Publisher:
Südwestdeutscher Verlag für Hochschulschriften
is a trademark of
Dodo Books Indian Ocean Ltd., member of the OmniScriptum S.R.L Publishing group
str. A.Russo 15, of. 61, Chisinau-2068, Republic of Moldova Europe
Printed at: see last page
ISBN: 978-3-8381-2394-3

Zugl. / Approved by: Hannover, Leibniz Universität Hannover, Diss., 2010

Copyright © Arman Mohseni
Copyright © 2011 Dodo Books Indian Ocean Ltd., member of the OmniScriptum S.R.L Publishing group

To my parents
Tooran and Mohammad Hassan
and my sisters and brother
Mahzad, Mahsa, and Ehsan

Abstract

Particle image velocimetry (PIV) is a non-intrusive flow measurement technique, capable of capturing a velocity field in either a series of single independent instantaneous measurements (conventional PIV), or a sequence of instantaneous measurements (high-speed PIV). It facilitates qualitative and quantitative study of flow structures in a plane or in a volume (tomographic PIV). PIV is a measurement technique of intense research interest to those investigating turbomachinery. In most of turbomachinery applications, the complexity of the flow structures requires the use of three-component or stereoscopic PIV (SPIV) and the geometrical scales of the machines require the use of endoscopes in order to access the flow channel. Compared to the traditional PIV, the introduced image distortion in the endoscopic SPIV is of higher order and is a source of error in the measurements. In the present work, an endoscopic SPIV setup is developed for flow investigations at the inlet of a process centrifugal compressor. Based on partial differential equations, a new image reconstruction method is presented, which is capable of resolving high order distortions of the endoscopic imaging. In order to improve the measurement accuracy, a method for the calibration of pressure and temperature signals under quasi-steady conditions for the existing measurement system is developed. Finally, several aspects of the application of the SPIV at the inlet of the compressor together with a typical result are presented for the first time.

Zusammenfassung

Particle Image Velocimetry (PIV) ist eine störungsfreie Messmethode, die es ermöglicht, das Geschwindigkeitsfeld entweder in einer Folge von unabhängigen momentanen Aufnahmen (konventionelles PIV) oder eine Reihe von momentanen Aufnahmen (High-Speed-PIV) zu erfassen. Es ermöglicht die qualitative und quantitative Untersuchung der Strömungsstrukturen in einer Ebene oder in einem Volumen (tomographisches PIV). PIV ist von intensivem Forschungsinteresse in Turbomaschinuntersuchungen. In meisten Anwendungen in Turbomaschinen die Komplexität der Bauart erfordert den Einsatz von drei-Komponenten- oder stereoskopischen PIV (SPIV). Die geometrischen Größen der Maschinen erfordern den Einsatz von Endoskopen, um den Strömungskanal zu erreichen. Im Vergleich zum traditionellen PIV, ist die eingeführte Bildverzerrung in endoskopischen Aufnahen von höherer Ordnung und ist eine Quelle der Messungenauigkeit. In der vorliegenden Arbeit wird ein endoskopischer SPIV-Aufbau für die Stömungsuntersuchungen am Eintritt eines Radialverdichters entwickelt. Basiert auf partiellen Differenzialgleichungen, wird ein neues Bildrekonstruktionsverfahren für die Datenverarbeitung hochverzerrten Bildern vorgestellt. Um die Messgenauigkeit zu verbessern, wird ein Verfahren für die Kalibrierung von Druck- und Temperatursignale unter quasi-stationären Bedingungen für das bestehende Messsystem entwickelt. Schließlich werden erstmals einige Aspekte der Anwendung von SPIV am Eintritt des Radialverdichters zusammen mit einem typischen Ergebnis dargestellt.

Acknowledgments

I would like to express my best thanks to Prof. Dr.-Ing. Jörg R. Seume for providing me the opportunity of research work at the Institute of Turbomachinery and Fluid Dynamics (TFD) and entering me to the beautiful world of flow measurements and the challenging field of turbomachinery. I gratefully appreciate his continuous support and valuable guidance during the development of this work. I wish to express my appreciation and thanks to Prof. Dr.-Ing. habil. Dr.-Ing. E.h. Dr. h.c. Friedrich-Wilhelm Bach for chairing the thesis committee and Prof. Dr. Friedrich Dinkelacker for accepting the evaluation of this research work and also to Prof. Dr.-Ing. habil. Markus Raffel for accepting the evaluation of this work and for his support and helpful advice during the development of the PIV setup for the measurements on the compressor test facility.

I wish to express my thanks and gratitude to the following people, whose support has made this research work possible and has helped the enhancement of its quality in several ways:

Dr.-Ing. Joachim Runkel and Dr.-Ing. Yavuz Gündogdu as senior engineers for their support and guidance especially during the experimental part of this work; Dr.-Ing. Erik Goldhahn for the informative and detailed discussions on the PIV measurement technique and for the introduction to the compressor test facility; Dipl.-Ing. Jasper Kammeyer, as group leader, and Dipl.-Ing. Michael Bartelt for their wonderful cooperation and help during the experimental part of this research work; my colleagues at TFD for their support and for providing a friendly atmosphere, especially, my colleagues in the secretariat and accountancy for their kind, sincere, and perfect support; the technical staff and different disciplines in the workshop including manufacturing, electronics, sensors, mechanical, and electrical at TFD for detailed discussions, informative talks, helpful opinions and advice, and effective cooperation during the design, construction, installation, and measurements; Mr. Raimund Fraatz, Mr. Johannes Peter, Mr. Tore Fischer, and Mr. Jan Terelak for their help during the installation, tests, and calibration of the measurement system; my family, to whom this research work is dedicated, my relatives, and friends for all of their support and encouragement.

I wish to express my intense gratitude to the German Academic Exchange Service (DAAD[1]) for the funding of this research work over the past four years and would like to express my thanks to the DAAD staff for their perfect support.

---

[1]DAAD: Deutscher Akademischer Austausch Dienst

Contents

Abbreviations	**11**
List of Figures	**13**
List of Tables	**17**
List of Symbols	**19**
Mathematical Definitions	**23**
1 Introduction	**25**
2 Distortion Compensation of Digital Images	**29**
2.1 Introduction	29
2.1.1 Calibration grid	30
2.1.2 Grid recognition in digital images	31
2.1.3 Refinement of node coordinates	34
2.1.4 Grid refinement	35
2.2 Back-transformation	38
2.2.1 Introduction	38
2.2.2 Mapping by the Poisson equation	40
2.2.3 Discretization	43
2.2.4 Computation of the source functions	44
2.2.5 Boundary and internal conditions	46
2.2.6 Solution of the Poisson equations	46
2.3 Image reconstruction	47
2.4 Applications	48
2.5 Conclusions	59
3 Theory of Local Measurements under Quasi-Steady Conditions	**61**
3.1 Introduction	61
3.2 Measurement	62

	3.3	Zero-drift of pressure sensors	64
	3.4	Calibration of local measurements	65
	3.5	Correlation functions	69
	3.6	Analysis of measured data	70
	3.7	Propagation of error	72
	3.8	Conclusions	74
4	**Velocity Measurement at the Compressor Inlet by Pressure Probes**		**75**
	4.1	The test facility	75
	4.2	Pressure and temperature measurement	75
	4.3	Zero-drift of pressure sensors	77
	4.4	Calibration of pressure and temperature measurement loops under quasi-steady conditions	81
	4.5	Velocity measurement by 5-hole pressure probes	81
	4.6	Results of velocity measurement by pressure probes	86
5	**PIV Measurement at the Compressor Inlet**		**93**
	5.1	Endoscopic SPIV setup at the compressor inlet	93
	5.2	Calibration	94
	5.3	Measurements	97
	5.4	Data analysis and results	103
6	**Conclusions and Future Work**		**109**
Bibliography			**111**
A	**Polynomial Fit with the Method of Least Squares**		**115**
B	**Spline Interpolation**		**117**
C	**Gaussian Function**		**119**
D	**Nonlinear Function Fitting**		**123**
E	**Discretization**		**129**
F	**Calibration Correlations of 5-hole Pressure Probes**		**131**
G	**Calibration Charts**		**137**

Abbreviations

amb	Ambient
CCD	Charge coupled device
DAAD	Deutscher Akademischer Austausch Dienst
DEHS	Di-ethyl-hexyl-sebacate
emf	electromotive force
FDM	Finite difference method
HWA	Hot-wire anemometry
IGV	Inlet guide vane
L2F	Laser two-focus velocimetry
LDA	Laser Doppler anemometry
LDV	Laser Doppler velocimetry
LTV	Laser transit velocimetry
Nd:YAG	Neodymium Yttrium Aluminum Garnet
PDE	Partial differential equation
PIV	Particle image velocimetry
SPIV	Stereoscopic particle image velocimetry
TFD	Institute of Turbomachinery and Fluid Dynamics
Typ.	Typical

List of Figures

2.1	Artificial Gaussian patterns as grid points in a calibration image	31
2.2	The image plane with the nomenclature and the coordinates of a digital image	32
2.3	Identification of nodes in the digital image of a distorted grid	33
2.4	Triangle with vector definitions	36
2.5	Planar convex quadrilateral	37
2.6	Grid refinement	38
2.7	Back-transformation	42
2.8	Discretization patterns for the rectangular grid of domain D_ξ	43
2.9	Interpolation	44
2.10	Interpolation along the common side of two adjacent cells	45
2.11	Application of the boundary conditions	46
2.12	Deformation of pixels after reconstruction	48
2.13	Synthetic grid from smooth curves	49
2.14	**(a)** the original grid (black) and the refined grid (gray) after grid identification and the solution of the Poisson equations on the grid of Fig. 2.13 and **(b)** the distribution of the distances between the nodes in the original grid and their corresponding nodes in the refined grid	50
2.15	The distribution of the source functions in the image plane	51
2.16	The distribution of the source functions in the object plane	51
2.17	**(a)** the coordinate lines after transformation and **(b)** the reconstructed grid	52
2.18	Synthetic grid from smooth curves	52
2.19	**(a)** the original grid (black) and the refined grid (gray) after grid identification and the solution of the Poisson equations on the grid of Fig. 2.18 and **(b)** the distribution of the distances between the nodes in the original grid and their corresponding nodes in the refined grid	53
2.20	The distribution of the source functions in the image plane	54
2.21	Reconstructed grid	54
2.22	A synthetic grid, generated by uniform distortion of a rectangular grid	55

2.23 (a) the original grid (black) and the refined grid (gray) after grid identification and the solution of the Poisson equations on the grid of Fig. 2.22 and (b) the distribution of the distances between the nodes in the original grid and their corresponding nodes in the refined grid 56
2.24 The distribution of the source functions in the image plane 57
2.25 The distribution of the source functions after transformation 58
2.26 (a) the coordinate lines after transformation and (b) the reconstructed grid 60

3.1 Typical zero-drifts of five pressure channels measured concurrently during 66 hours . 65
3.2 A typical calibration sample of a pressure transmitter 67
3.3 The calibration correlation function and the dependency of the uncertainty parameters 71

4.1 The simplified schematic diagram of the compressor with the IGV 76
4.2 The process flow diagram of the compressor test facility, with valves positioned for the open-loop operation 77
4.3 The measurement system diagram 78
4.4 The zero-drift of the pressure transducers during one day (channels 201 to 213) . . . 79
4.5 The zero-drift of the pressure transducers during one day (channels 301 to 313) . . . 80
4.6 The calibration setup diagrams 82
4.7 The schematic horizontal section of the compressor through the impeller center line . 83
4.8 The definition of the pressures and coordinates of the 5-hole pressure probe for measurement in the calibrated mode 84
4.9 Typical results of the steady state measurement of the primary quantities 88
4.10 The corresponding pressure values of Fig. 4.9 after data analysis 89
4.11 Error ranges corresponding to the pressure values of Fig. 4.10 90
4.12 The variation of the static and stagnation pressures in planes 1 and 2 in the spanwise direction 91
4.13 Velocity variations in planes 1, 2, and 3 in the spanwise direction 91
4.14 Variation of the Mach number in planes 1, 2, and 3 in the spanwise direction 92

5.1 The endoscopic stereoscopic PIV setup on the compressor for flow measurements between the IGV and the impeller 95
5.2 The PIV measurement section 96
5.3 (a) the calibration grid and its adjusting mechanism, (b) and (c) the calibration images of camera 1 and camera 2, respectively 98
5.4 A sub-domain of a calibration image of camera 1, before (left) and after (right) averaging 99
5.5 The dewarped calibration images of (a) camera 1, (b) camera 2, and (c) the recombined dewarped images of cameras 1 and 2 100
5.6 The distribution of the distances between the nodes in the original grid and their corresponding nodes in the refined grid (a) camera 1 and (b) camera 2 101

List of Figures

5.7   Comparison between **(a)** the reconstruction method presented in Ch. 2 and **(b)** reconstruction by a second order analytic projection 102

5.8   Typical PIV records **(a)** camera 1, **(b)** camera 2, **(c)** camera 1 dewarped, and **(d)** camera 2 dewarped 104

5.9   The rotation of the light-sheet, region A, from its expected direction, which is along the machine axis and parallel to the rectangle B, and the effect of droplet formation on the end window of the endoscope, region C 105

5.10 The location of the light-sheet reflection patterns after the image reconstruction for the case of misalignment between the light-sheet and the calibration plane 106

5.11 The projected average velocity field on the light-sheet plane 107

5.12 Comparison between the pressure probe and the PIV measurements 107

A.1   The vertical and perpendicular offsets of a point from a curve 115

E.1   The definitions of the discretization on a rectangular grid 129

G.1   The Barton-cell absolute pressure transmitter for the measurement of the ambient pressure 138

G.2   Pressure channel 301, corresponding to the pressure p_1 of the probe in plane 1 139

G.3   Pressure channel 302, corresponding to the pressure p_2 of the probe in plane 1 139

G.4   Pressure channel 303, corresponding to the pressure difference $p_2 - p_3$ of the probe in plane 1 140

G.5   Pressure channel 305, corresponding to the pressure difference $p_4 - p_5$ of the probe in plane 1 140

G.6   Pressure channel 306, corresponding to the pressure p_1 of the probe in plane 2 141

G.7   Pressure channel 307, corresponding to the pressure p_2 of the probe in plane 2 141

G.8   Pressure channel 308, corresponding to the pressure difference $p_2 - p_3$ of the probe in plane 2 142

G.9   Pressure channel 309, corresponding to the pressure difference $p_4 - p_5$ of the probe in plane 2 142

G.10 Pressure channel 310, corresponding to the pressure p_1 of the probe in plane 3 143

G.11 Pressure channel 311, corresponding to the pressure p_2 of the probe in plane 3 143

G.12 Pressure channel 312, corresponding to the pressure difference $p_2 - p_3$ of the probe in plane 3 144

G.13 Pressure channel 313, corresponding to the pressure difference $p_4 - p_5$ of the probe in plane 3 144

List of Tables

2.1	Summary of the set definitions used in the image reconstruction	40
3.1	Summary of the statistical analysis in the quasi-steady calibration	69
4.1	The characteristics and performance data of the compressor test facility	76
F.1	The coefficients of the pitch angle correlation, Eq. (F.7)	133
F.2	The coefficients of the dynamic pressure correlation, Eq. (F.8)	134
F.3	The coefficients of the static pressure correlation, Eq. (F.9)	135
F.4	The coefficients of the total pressure correlation, Eq. (F.10)	136
G.1	Channel assignments of the pressure probes	137
G.2	Maximum zero-drifts of the pressure channels during 24 hours	138

List of Symbols

Symbol	Unit	Definition
∂A		the boundary of a geometrical domain A
$C(O, r)$		circle C centered at O and with radius r
C_{DP}	[-]	Dynamic Pressure Coefficient, Eq. (4.10)
C_p	[J/(kg·K)]	specific heat at constant pressure
C_{PA}	[-]	Pitch Angle Coefficient, Eq. (4.7)
C_{SP}	[-]	Static Pressure Coefficient, Eq. (4.8)
C_{TP}	[-]	Stagnation Pressure Coefficient, Eq. (4.9)
C_{YA}	[-]	Yaw Angle Coefficient, Eq. (4.6)
D_1		the set of image pixels or the digital image in the image plane, Sect. 2.1.2
D_1'		the indexing set of D_1
D_2		the subset of D_1 of the maximum local intensities, Eq. (2.7)
D_2'		the indexing set of D_2
D_3		the set of pixels of a back-transformed image, Sect. 2.3
D_3'		the indexing set of D_3
d_1	[mm]	pixel spacing in a digital image in the y_1 direction, Fig. 2.2
d_2	[mm]	pixel spacing in a digital image in the y_2 direction, Fig. 2.2
f	[-]	probability density function
$f(x_i; a_j)$		a (mathematical) function of n variables $x_1, \ldots, x_n$ and m parameters $a_1, \ldots, a_m$
G_1		a rectangular Cartesian grid of points in the object plane, equivalent to Γ_1 in the image plane
G_1'		the indexing set of G_1
G_2		refined G_1 with uniform subdivision of cells, equivalent to Γ_2 in the image plane
G_2'		the indexing set of G_2
$\mathbb{I}$		the set of non-negative integers, $\mathbb{I} \equiv \mathbb{Z}^{\oplus}$
I_1		the set of the light intensities assigned to the D_1 elements
I_1'		the indexing set of I_1

Symbol	Unit	Definition
I_2		the set of the light intensities assigned to the D_3 elements
I'_2		the indexing set of I_2
i		index
J		the Jacobian
$\tilde{J}$		the inverse Jacobian, Eq. (2.31)
j		index
M	[-]	the Mach number
m		index, Fig. 2.2
$\mathbb{N}$		the set of natural numbers
n		index, Fig. 2.2; time step or iteration step, Eq. 2.39
p	[Pa]	absolute pressure
p_1 to p_5	[Pa]	the pressure values of a 5-hole pressure probe, Figs. 4.3 and 4.8a
$\bar{p}_{23}$	[Pa]	the arithmetic average of the probe pressures p_2 and p_3
$\overset{i}{p}$		the source function of the i-th Poisson equation in the image plane
$\overset{i}{q}$		the inverse source function of the i-th Poisson equation in the object plane
R	[J/(kg·K)]	gas constant (of air)
$\mathbb{R}$		the set of real numbers
r		temperature recovery factor, Eq. (4.13)
S_k		the set of neighbor pixels, Eq. (2.8)
T	[K]	absolute temperature
T_0	[K]	absolute stagnation temperature
T_R	[K]	absolute recovery temperature, Eq. (4.13)
T_{st}	[K]	absolute static temperature
t	[s]	time
V	[m/s]	the magnitude of the absolute flow velocity
V_1, V_2, V_3	[m/s]	the components of the absolute flow velocity in the $x_1 x_2 x_3$-coordinates in Fig. 4.8 at its reference position (i.e. $x'_1 x'_2 x'_3$)
x		the x coordinate of the velocity field measured by PIV, Fig. 5.11
$\boldsymbol{x}$		a geometric location in $\mathbb{R}^3$
x_i		the i-th component of the coordinates in the image plane, Fig. 2.2; the i-th component of the coordinates of a pressure probe
x'_i		the i-th component of the reference coordinates of a pressure probe, Fig. 4.8
y		the y coordinate of the velocity field measured by PIV, Fig. 5.11
$\mathbb{Z}$		the set of integers

Greek Letters:

α	[°]	yaw angle

List of Symbols

Symbol	Unit	Definition
α'	[°]	yaw angle in the reference coordinates $x'_1 x'_2 x'_3$
α_0	[°]	the offset of the yaw angle
β_i		the parameters of the Gaussian function, Eq. (2.10)
Γ_0		the grid of the approximate locations of the Gaussian patterns in the image plane as described in Sect. 2.1.3
Γ'_0		the indexing set of Γ_0
Γ_1		the distorted grid in the image plane as described in Sect. 2.1.3, equivalent to G_1 in the object plane
Γ'_1		the indexing set of Γ_1
Γ_2		refined Γ_1 as described in Sect. 2.1.4 with the same number of cell subdivisions as in G_2, equivalent to G_2 in the object plane
Γ'_2		the indexing set of Γ_2
γ	[°]	pitch angle
γ'	[°]	pitch angle in the reference coordinates $x'_1 x'_2 x'_3$
$\check{\Delta}(\tilde{\psi})$	†	the minimum overall deviation of a variable ψ from its mean
$\hat{\Delta}(\tilde{\psi})$	†	the maximum overall deviation of a variable ψ from its mean
δ_0	†	zero-drift
$\check{\delta}_1$	†	minimum device internal error
$\hat{\delta}_1$	†	maximum device internal error
$\delta_{i,j}$	†	the distance between the original nodes and their corresponding nodes in the reference grid, Eq. (2.44)
$\check{\epsilon}$	†	minimum of sample
$\hat{\epsilon}$	†	maximum of sample
κ	[-]	isentropic exponent
ξ_i	[mm]	the i-th component of coordinates in the object plane
φ	†	a sample value in a reference measurement system
Ψ^d	†	the range of the variation of a digital quantity, Eq. (3.6)
ψ	†	a sample value in a measurement system
σ	†	standard deviation

Subscripts:

$*$		scaled variable, App. D
0		stagnation conditions
i		the index of a data-point
j		the index of a sample member
r		reference parameter/variable
st		static conditions

Symbol	Unit	Definition

Superscripts:
d — a discrete variable or quantity
T — matrix transposition

Over sets:
$\bar{a}$ — the special time average of a
$\bar{a}'$ — signal noise, based on the special time average of a
$\check{a}$ — the cumulative time average of a
$\tilde{a}$ — the estimated time average or the mean of a
$\tilde{a}'$ — signal noise, based on the estimated time average or the mean of a
$\vec{a}$ — vector a
$\hat{a}$ — unit vector a; the maximum of a sample
$\overset{k}{x}$ — the k-th element of the set $\{\overset{k}{x}\}$

Mathematical Symbols:
$\emptyset$ — the empty set
$|$ or $:$ — ... such that ...; ... it is true that ...
$|$ — separator between a discrete quantity and its indices
$|\vec{a}|$ — the absolute value or the length of a vector $\vec{a}$
$\equiv$ — equivalent to; identically equal to
$\times$ — vector cross product
$:=$ — defined as; equal to by definition
$\forall$ — for all elements
$\because$ — because of
$\therefore$ — therefore; it follows that
$A \Rightarrow B$ — therefore; it follows that; implication; if A then B
$A \Leftrightarrow B$ — equivalence; A if and only if B
$f : \mathrm{D} \mapsto \mathrm{R}$ — function or mapping f from the domain set D into the range set R
$\square$ — Q.E.D. symbol; the end of a definition, theorem, proof, ...

† The unit varies according to the corresponding (physical) quantity.

Mathematical Definitions

Bold face letters are used for the variables of dimensions higher than zero, such as vectors and tensors, as well as matrices.

The following definitions are used throughout this work:

Partial Derivatives

Partial derivatives are alternatively denoted by a comma followed by indices in the subscript. For an arbitrary function $\varphi_k(x_1, x_2, x_3)$:

$$\varphi_{k,m}(x_1, x_2, x_3) := \frac{\partial \varphi_k(x_1, x_2, x_3)}{\partial x_m}$$

$$\varphi_{k,mn}(x_1, x_2, x_3) := \frac{\partial^2 \varphi_k(x_1, x_2, x_3)}{\partial x_m \partial x_n}$$

Sets

For a given set of numbers A:

$$A^+ := \{x \in A \mid x > 0\}$$
$$A^- := \{x \in A \mid x < 0\}$$
$$A^\oplus := \{x \in A \mid x \geq 0\}$$
$$A^\ominus := \{x \in A \mid x \leq 0\}$$

Intervals

The intervals, defined on a set of numbers A, are as follows:

$$[a, b]_A := \{x \in A \mid a \leq x \leq b\}$$
$$]a, b[_A \equiv (a, b)_A := \{x \in A \mid a < x < b\}$$

$$[a,b[_A \equiv [a,b)_A := \{x \in A|\ a \leq x < b\}$$
$$]a,b]_A \equiv (a,b]_A := \{x \in A|\ a < x \leq b\}$$

If A is omitted, then $\mathbb{R}$ is assumed.

Functions

Bracket function
On a finite or countable set of numbers A^d,

$$A^d = \{f(k)|\ f(k) : \mathbb{Z} \longmapsto \mathbb{R},\ f(k) \text{ monotone and increasing}\}$$

the bracket of a real variable x is defined as:

$$[x]_{A^d} : \mathbb{R} \longmapsto A^d$$
$$[x]_{A^d} = f(k) \iff \exists k \in \mathbb{Z}:\ f(k) \leq x < f(k+1)$$

Conditional count function
On a sample $\{x_i\}$, the *conditional count*, $\mathcal{N}_{N_1}^{N_2}(x_i\ |\ A)$, is equal to the number of x_i, $i \in [N_1, N_2]_{\mathbb{Z}}$, for which a logical condition or a set of logical conditions A is true.

Min and max functions
For a real function $f(x)$:

$$y = \min_A (f(x)) \iff \{\forall x \in A:\ f(x) \geq y\}$$
$$y = \max_A (f(x)) \iff \{\forall x \in A:\ f(x) \leq y\}$$

Order of magnitude

The order of magnitude is represented either in its formal form $O(x^m, y^n)$, standing for all terms of degree m or higher with respect to x and of degree n or higher with respect to y, or as $O^n(x_1, \ldots, x_k)$, which stands for all terms of degree n or higher with respect to x_1 to x_k.

Chapter 1

Introduction

Particle Image Velocimetry (PIV) is the outcome of scientific research and technological developments in optics, digital imaging, laser technology, electronics, and computer sciences. During the past 20 years, these improvements have brought major advances in this measurement technique and still further developments need to be made in order to find solutions for current short comings of the method.

As a non-intrusive flow measurement technique, capable of providing a planar representation of a flow field, PIV has become a method of choice for the flow investigations in turbomachines. Compared to the single-point measurement techniques such as HWA[1], LTV[2], LDA[3], and multi-hole pressure probes, this planar representation of instantaneous velocity field in successive time-independent measurements (traditional PIV) as well as phase-resolved measurements (high-speed PIV) reveals the flow structures and can be used to investigate the kinematics of a flow. In this regard, PIV is of intense interest for investigating the fluid flow phenomena in turbomachines. An overview of the recent applications of PIV in thermal turbomachinery can be found in Woisetschläger and Göttlich (2008).

The application of PIV in turbomachines, as compared with its applications in external flows or large scale internal flows, is rather challenging. In most machines, the geometrical complexity limits the accessibility of the flow channel, especially if the nature of the flow requires three-component velocity measurements. Light reflection from the machine components is another issue, which determines the extents of the measurable range. The limitations due to light reflection and background illumination are more severe, if oil droplets are used for seeding, which have considerably lower illumination than solid particles.

The calibration of a PIV setup often requires the imaging of a geometrically known object in the prospective measurement section. In the PIV setups, which involve image distortion, such as the stereoscopic setups, the reference object should cover the whole measurement section. In such cases, the lack of accessibility of the flow channel may require external calibration.

---

[1] HWA: Hot-wire anemometry
[2] LTV: Laser transit velocimetry. Also known as L2F: Laser two-focus velocimetry
[3] LDA also LDV: Laser Doppler anemometry/velocimetry

If the measurement section can be accessed through a window, so that the light-sheet can be captured by objectives, the PIV images will be of higher quality. In such measurements, which are generally suitable for large-scale machines, the light-sheet images could be guided directly (Voges et al., 2007; Liu et al., 2006) or indirectly via mirror deflections (Wernet et al., 2001; Zachau et al., 2008) to camera. More flexibility is achieved by using an endoscope for light-sheet delivery into the measurement section, without substantial reduction in the quality of the measurements (Yun et al., 2008).

The implementation of endoscopes both for camera(s) and light-sheet delivery, known as *endoscopic PIV*, provides more accessibility and covers a broader range of applications. The endoscopic PIV is not only a solution for small-scale measurements, but also provides enough flexibility to achieve feasible configurations. Some of its drawbacks are the reduction of the intensity and the quality of the light-sheet, decreased signal-to-noise ratio in the recordings, and image distortions of higher orders than that of imaging by objectives and light-sheet optics. In this regard, the endoscopic PIV is more sensitive to the settings and the quality of a PIV setup, than the traditional PIV. While the light-sheet and the image quality can be optimized by improving the settings and the quality of the measurement setup and section, the image distortion should be corrected during data analysis.

The present research work concerns the development of the PIV measurement technique for flow investigation at the inlet of a centrifugal compressor with inlet guide vanes (IGV). The current applications of the endoscopic PIV in turbomachinery consist of two-component measurements, in which the optical axis of the camera is perpendicular to the light-sheet (Kegalj and Schiffer, 2009; Dierksheide et al., 2002). In order to capture the three components of the flow velocity, the endoscopic stereoscopic PIV is employed for the first time at the compressor inlet. Two camera endoscopes together with a light-sheet endoscope provide a three-component PIV setup in order to investigate the flow structure between the IGV and the impeller.

One characteristic of the stereoscopic endoscopic imaging is highly distorted recordings. Image distortion is a common issue in the most of PIV applications. In imaging by objectives, the effect of image distortion may be neglected, if the view axis is normal to the object plane. In stereoscopic imaging with objectives, usually a first-order mapping is sufficient for the distortion compensation of the images. Methods to compensate for higher order distortions have been developed by implementing analytic mapping functions in camera models (Coudert and Schon, 2001; Soloff et al., 1997). Although the available reconstruction methods are suitable for objective recordings, they are incapable of full reconstruction of endoscopic measurements. The issue of image distortion is considered in Ch. 2. A novel method for image reconstruction based on partial differential equations is presented, which is capable of full reconstruction of the endoscopic recordings.

Besides PIV measurements, the inlet flow is investigated with the aerodynamic pressure probes. In regular on-site calibration procedures, the disturbances due to small instabilities in the device and ambient conditions are ignored. This may result in the extended ranges of uncertainty. In order to enhance the accuracy of the measurements, the issue of on-site calibration has been considered in detail. A theory for the calibration under quasi-steady conditions has been developed in Ch. 3. By inclusion of low rate variations in the calibration data, the theory provides means to improve the accuracy of the calibration. This chapter provides the theoretical basis for the data analysis of the flow measurements by the pressure probes, which are presented in Ch. 4.

Chapter 5 presents the first flow investigation by SPIV at the compressor inlet. The calibration of the stereoscopic setup and the application of the image reconstruction method of Ch. 2 is presented in Sect. 5.2. In this chapter the achieved state of the endoscopic SPIV approach is shown, which was reached from the first few measurements in the compressor. The practical features of the application of the endoscopic SPIV for the flow investigation at the compressor inlet are discussed and are followed by the conclusions and necessary future work in Ch. 6.

Chapter 2

Distortion Compensation of Digital Images

2.1 Introduction

In stereoscopic setups the optical axis of the imaging system is oblique to the object plane and image sharpness is achieved according to the Scheimpflug criterion (Scheimpflug, 1904). Oblique angles of view introduce image distortion, which is linear in pin-hole camera models. Another type of distortion is due to the lens quality and the imperfections of the optical components, which can be of second or higher orders.

As a result of distortion, the length scale of an image is altered. The variation of the length scale is a property of the imaging system and is, in general, unknown. The distances recorded in an image, whose length scale is not constant, cannot be measured directly and in order to extract quantitative information, its distortion should be identified or corrected first.

One method of distortion compensation is the use of analytical mappings by polynomials or the ratio of polynomials, Raffel et al. (2007). In this method, the image of a known reference grid, such as a grid of point patterns, is used to extract the distortion data of the imaging system. An analytical function is fitted on the data so that the distorted image be mapped into the reference grid. The mapping is valid for all image records of the same imaging system at the same configuration and can be applied to PIV images for back transformation.

Another approach is the successive transformations of an artificially created reference image (Baggenstoss, 2004). In an iterative procedure, several distorted reference images are compared with a test image by correlation, until a best match is obtained. The method has high computational burden at preprocessing, which includes a training phase, but requires very low computation, when determining the distortion of the test images. The applications of this method include pattern recognition in satellite and medical imaging.

By evaluating the image position changes due to small sample position shifts, Pollak et al. (2001)

present a novel method of correcting image distortion. This method does not require a sample (calibration) image and uses the identifiable features of the distorted image. Three position shifts of an object are recorded and the displacements of image features are used to provide a polynomial fit to the distortion field.

In most single medium stereoscopic imaging setups, such as imaging with objectives or near field endoscopic imaging, the use of polynomial fits for distortion compensation provides enough accuracy. However, high order or local distortions require more sophisticated mapping functions. Partial differential equations (PDEs) have found applications in many image processing methods such as image inpainting, dejittering, registration, and segmentation (Tai et al., 2007). A good survey on different PDE models used in tomographic image reconstruction methods has been presented by Natterer (2006).

In this chapter, a novel PDE-based distortion compensation method is presented. The method covers a broad class of image distortions, including local distortions, and can be used for back-projection of endoscopic PIV records. The method consists of three main parts:

1. Grid recognition in digital images
2. Back-transformation
3. Image reconstruction

which are presented in the following.

2.1.1 Calibration grid

One common method for determining the distortion of an image is the comparison of the image records of geometrically definite planar patterns with themselves. A rectangular grid of dots or crosses, a grid of horizontal and vertical lines, and a chessboard pattern are some commonly used examples. A pattern may be carved on a metal sheet by high precision machining or may be printed by high resolution printer devices. The pattern is then recorded by an imaging system. The resulting image contains distortion information of the imaging system.

The physical dimensions of a pattern can be determined by a direct measurement within the accuracy of the measurement system and the measurement method. In a distorted image, the length scale is usually not constant. Therefore, the lengths in a distorted image cannot be determined directly.

Digital imaging confines the resolution of an image to the effective pixels of the recording elements, such as a CCD[1] chip. A sub-pixel accuracy, however, is achievable by interpolating pixel values. Electronic cameras also add noise to the images. Averaging techniques can be used to suppress the noise effects. Digitization and addition of noise introduce additional image distortions, which are independent of the optics of the imaging system.

---

[1]CCD: Charge Coupled Device

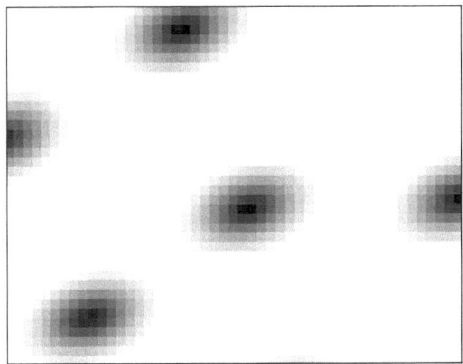

Fig. 2.1: Artificial Gaussian patterns as grid points in a calibration image
The image is inverted for printing.

One method for compensating noise effects and achieving sub-pixel accuracy in pattern recognition in a digital image, is the use of definite continuous gray scale patterns and interpolating functions. Figure 2.1 shows an artificial Gaussian pattern around a central point. Interpolation of the image of the pattern with the Gaussian function results in the coordinates of the central point with sub-pixel accuracy. Owed to the random nature of the image noise, the Gaussian fit on several pixels can reduce the noise effects. In a comparison between three center estimation methods, namely the binary center of mass (barycenter), the gray scale center of mass, and the Gaussian fit, Udrea et al. (2000) show the superior accuracy of the Gaussian fit method.

In this work, a rectangular grid is used as the calibration grid. The grid points are marked by symmetric gray scale Gaussian patters centered at them. In the following, a method for grid recognition in distorted images is presented. The method identifies the location of the grid nodes with pixel accuracy. The Gaussian fit is then used on the point patterns to refine the node coordinates to sub-pixel accuracy.

2.1.2 Grid recognition in digital images

A digital gray scale image[1] is considered as a simply connected rectangular closed domain D_1 bounded by ∂D_1, Fig. 2.2, and composed of an array of $(m_{max} - m_{min} + 1) \times (n_{max} - n_{min} + 1)$ simply connected open sub-domains $D_1|_{m,n}$, each of which bounded by a boundary $\partial D_1|_{m,n}$, which satisfy the following conditions:

$$\mathbf{M} := [m_{min}, m_{max}]_{\mathbb{Z}} \tag{2.1}$$

---

[1]In this work only gray scale digital images are considered. Without loss of generality, the black color is assumed for the lowest light intensity and the white for the highest.

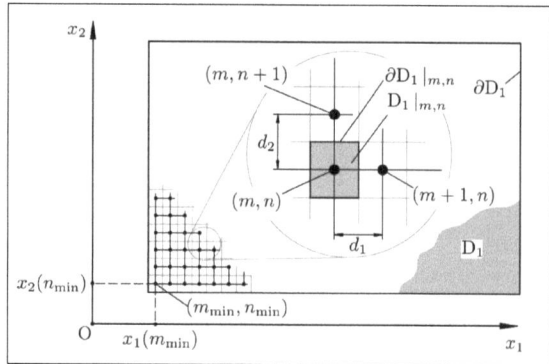

Fig. 2.2: The image plane with the nomenclature and the coordinates of a digital image

$$\mathbf{N} := [n_{\min}, n_{\max}]_{\mathbb{Z}} \tag{2.2}$$

$\forall m_1, m_2 \in \mathbf{M} \land \forall n_1, n_2 \in \mathbf{N}:$

$$m_1 \neq m_2 \lor n_1 \neq n_2 \Rightarrow \mathbf{D}_1|_{m_1,n_1} \bigcap \mathbf{D}_1|_{m_2,n_2} = \emptyset \tag{2.3}$$

$$\mathbf{D}_1 = \bigcup_{\substack{m \in \mathbf{M} \\ n \in \mathbf{N}}} (\mathbf{D}_1|_{m,n} \cup \partial \mathbf{D}_1|_{m,n}) \tag{2.4}$$

It is assumed that the sub-domains are equal rectangles and the location of each sub-domain can be described with respect to its centroid in the Cartesian coordinates Ox_1x_2, Fig. 2.2:

$$x_1(m) = x_1(m_{\min}) + d_1(m - m_{\min}) \tag{2.5}$$
$$x_2(n) = x_2(n_{\min}) + d_2(n - n_{\min}) \tag{2.6}$$

In a digital image, each sub-domain $\mathbf{D}_1|_{m,n}$, called a *pixel*, is assigned a discrete light intensity value $I_1|_{m,n} \in \mathbb{I}$, zero for the black color and maximum for the white color. This value is assigned to the centroid of each pixel.

Figure 2.3 shows a digital image, domain $\mathbf{D}_1$, of a distorted rectangular grid of the Gaussian patterns in the x_1x_2 plane. The grid adapts the curvilinear coordinates $\xi_1\xi_2$. It is assumed that the Gaussian patterns are white patterns on a black background.

As the first step in grid identification, a subset of the image containing the local maxima of the intensities is extracted:

$$\mathbf{D}_2 = \{\mathbf{D}_1|_{m,n} \in \mathbf{D}_1 : \forall \mathbf{D}_1|_{p,q} \in S_k|_{m,n} \Rightarrow I_1|_{m,n} \geq I_1|_{p,q} > 0\} \tag{2.7}$$

Chapter 2. Distortion Compensation of Digital Images

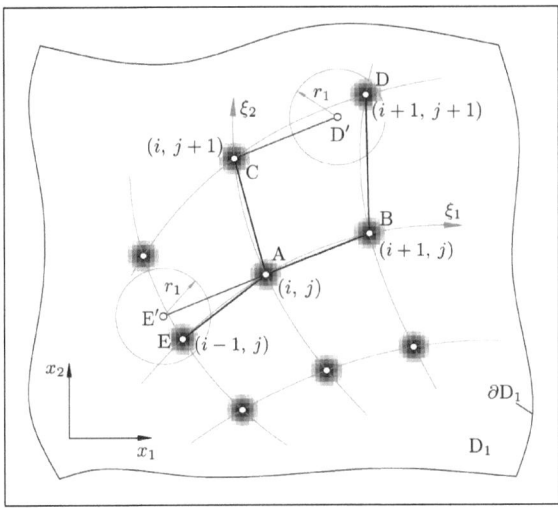

Fig. 2.3: Identification of nodes in the digital image of a distorted grid
Each node is identified by the maximum intensity in a Gaussian pattern. The image of Gaussian patterns is inverted for printing.

where

$$S_k |_{m,n} = \{D_1 |_{p,q} \in D_1 : k \in \mathbb{N}, \; p \in [m-k, \; m+k]_\mathbb{Z}, \\ q \in [n-k, \; n+k]_\mathbb{Z}\} \tag{2.8}$$

In order to treat the boundary pixels, D_1 may be extended by zero-padding or $S_k |_{m,n}$ can be modified in the neighborhood of the boundary to include the inner elements of D_1. The coordinates of the distinct pixels of D_2 reveal the approximate locations of the grid nodes. If the image is not saturated, D_2 should not contain multiple pixels in a Gaussian pattern. However, if the maximum of a Gaussian pattern is located on a pixel boundary, it can contain adjacent pixels with the same intensities. This can also happen due to the noise added by the imaging system, which alters the pixel intensities in a random manner. Since adjacent pixels are assumed to be in the same Gaussian pattern, only one set of coordinates is assigned to them. One way is to determine their common centroid. The set of the locations of the D_2 elements, i.e. the coordinates of the centroids of single elements or the common centroid of adjacent elements, is an approximation to the grid in the image plane and is referred to as $\Gamma_0 \subset \mathbb{R}^2$.

For the second step of grid recognition, the coordinate system of the grid should be known. A set of three points such as points A, B, and C in Fig. 2.3, indicating the axis directions, and a point as the origin of the coordinates $\xi_1 \xi_2$ should be given.

From A and B, E' is determined by $\overrightarrow{BA}$. A search on Γ_0 elements within $C(E', r_1)$ determines E as the nearest point to E'. The radius r_1 is selected proportional to $|\overrightarrow{BA}|$. Similarly $\overrightarrow{AB}$ at C results in D' and D is found by searching within $C(D', r_1)$. All other eight neighbors of A are found in similar ways. By repeating the same procedure on newly found nodes, the whole grid is captured. By registering the relation of each found node, the connectivity data of the grid points is determined during the grid identification.

2.1.3 Refinement of node coordinates

In order to achieve sub-pixel accuracy for the node coordinates, the Gaussian patterns in the image of the calibration grid are interpolated with the Gaussian function. This method reduces the effect of image noise and partly recovers the distortion of the Gaussian patterns caused by discretization in the digital image.

Given the position of a Gaussian pattern by its maximum intensity in the pixel range, a subset of M pixels including the maximum and its neighbor nodes is selected:

$$\left\{ (\overset{k}{x}_1, \overset{k}{x}_2, \overset{k}{y}_0) \right\}, \quad k \in [1, M]_{\mathbb{N}} \tag{2.9}$$

The Gaussian function:

$$\begin{aligned} y &= f(x_1, x_2; \beta_1, \ldots, \beta_6) \\ &= \exp(\beta_1 + \beta_2 \, x_1 + \beta_3 \, x_1^2 + \beta_4 \, x_2 + \beta_5 \, x_1 \, x_2 + \beta_6 \, x_2^2) \end{aligned} \tag{2.10}$$

with six parameters β_i, is fitted on the subset by nonlinear least squares fitting using the Levenberg-Marquardt algorithm. The details of the Gaussian function and the nomenclature used is given in App. C.

With the selected set of points and with the initial values of the parameters, the following matrices are calculated:

$$\boldsymbol{A} = [A_{mn}]_{M \times 6} \tag{2.11}$$
$$A_{m,1} = \overset{m}{y} \tag{2.11a}$$
$$A_{m,2} = \overset{m}{x}_1 \overset{m}{y} \tag{2.11b}$$
$$A_{m,3} = (\overset{m}{x}_1)^2 \overset{m}{y} \tag{2.11c}$$
$$A_{m,4} = \overset{m}{x}_2 \overset{m}{y} \tag{2.11d}$$
$$A_{m,5} = \overset{m}{x}_1 \overset{m}{x}_2 \overset{m}{y} \tag{2.11e}$$

$$A_{m,6} = (\overset{m}{x}_2)^2 \overset{m}{y} \tag{2.11f}$$

$$\boldsymbol{B} = \boldsymbol{A}^T \cdot \boldsymbol{A} \tag{2.12}$$

$$\boldsymbol{C} = [C_m]_{6\times 1}, \quad \boldsymbol{C} = \boldsymbol{A}^T \cdot \boldsymbol{E} \tag{2.13}$$

$$\boldsymbol{E} = [E_m]_{M\times 1}, \quad E_m = \overset{m}{y}_0 - \overset{m}{y} \tag{2.14}$$

where $m \in [1, M]_\mathbb{N}$ and $\overset{m}{y}$ is the value of Eq. (2.10) at $(\overset{m}{x}_1, \overset{m}{x}_2)$. The optimum set of parameters β_i, which minimizes the target function $\tilde{R}$, Eqs. (D.7) and (D.17), is the solution of

$$(\boldsymbol{B}^* + \lambda \boldsymbol{I}^*) \cdot \boldsymbol{D}_0^* = \boldsymbol{C}^* \tag{2.15}$$

for $\boldsymbol{D}_0^*$ by iteration. A star (*) as the superscript of a matrix, denotes an scaled matrix (see App. D).

The refinement of the node coordinates in Γ_0 yields a grid Γ_1 in $\mathbb{R}^2$ with one-to-one correspondence with the image of Gaussian patterns in D_1. It is assumed that the imaging system does not alter the topology of the object plane during imaging. Therefore, the image grid Γ_1 corresponding to a rectangular structured grid in the object plane will be structured and the grid coordinates ξ_1 ξ_2, Fig. 2.3, will be curvilinear in general.

In a typical structured grid Γ_1, the elements are called *grid points* or *nodes*. The curves $\xi_i = const.$, $i \in \{1, 2\}$, passing through the nodes are called *grid lines*. Grid lines can be indexed with successive integers. This in turn results in node indexing. Corresponding to Γ_1, an indexing set $\Gamma_1' \subset \mathbb{Z}^2$ is defined, whose members are in one-to-one correspondence to the members of Γ_1. It is assumed that the index values along a grid line increase in the direction of the coordinate axis with unit steps between the neighbor nodes, Fig. 2.3. A quadrilateral $\{(i, j), (i+1, j), (i+1, j+1), (i, j+1)\}$ is called a *(grid) cell*.

2.1.4 Grid refinement

The grid Γ_1 of the calibration image contains the distortion information of the imaging system. The back-transformation method in the following section uses the node locations in Γ_1 to determine a discrete transformation function, which results in a transformation data set. The data set is then used to back-transform the distorted images using interpolation. In order to enhance the accuracy of the back-transformation, especially for coarse grids, Γ_1 can be refined by subdividing its cells.

Grid refinement consists of the subdivision of grid cells in the object plane, the extension of the grid data-base to include the subdivided cells, and the initialization of the extended grid. The discretization of the Poisson equations in the following sections is valid for a rectangular grid with equal grid spacing in each direction of the coordinates. To maintain this property, the only valid grid refinement is a uniform subdivision of cells. The number of subdivisions in each coordinate direction remains the

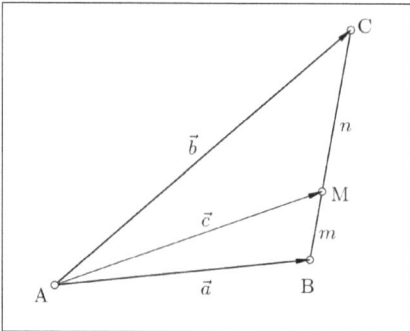

Fig. 2.4: Triangle with vector definitions

same for all cells but can differ between the coordinate directions.

The number of the subdivisions of the distorted grid in the image plane is the same as in the object plane. However, the coordinates of the added nodes are unknown and are determined by a solution of the transformation equations. An approximation of the coordinates of the added nodes can be used for initialization. The following theorems provide the tools for subdividing the distorted grid in the image plane.

Theorem 2.1 In a triangle ABC with a point M on BC and $m = $ BM and $n = $ CM, Fig. 2.4:

$$\vec{c} = \frac{\vec{a}+\vec{b}}{2} - \left(\frac{m-n}{m+n}\right)\frac{\vec{a}-\vec{b}}{2} \tag{2.16}$$

Theorem 2.2 In a planar convex quadrilateral ABCD, Fig. 2.5a, if the segments EF and GH split the sides so that:

$$\text{AE/EB} = \text{DF/FC} = \alpha \in \mathbb{R}^+ \quad \wedge \quad \text{AH/HD} = \text{BG/GC} = \beta \in \mathbb{R}^+ \tag{2.17}$$

then:

$$\text{HM/MG} = \alpha \quad \wedge \quad \text{EM/MF} = \beta \tag{2.18}$$

Proof: Figure 2.5b shows the quadrilateral with vector definitions. Assuming the points M_1 and M_2

Chapter 2. Distortion Compensation of Digital Images

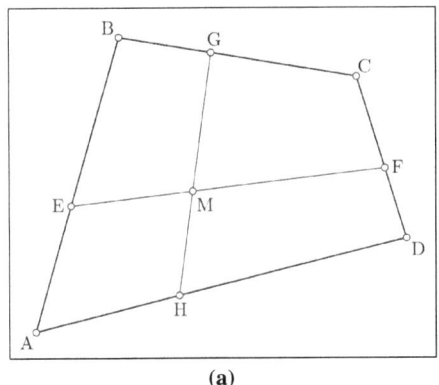
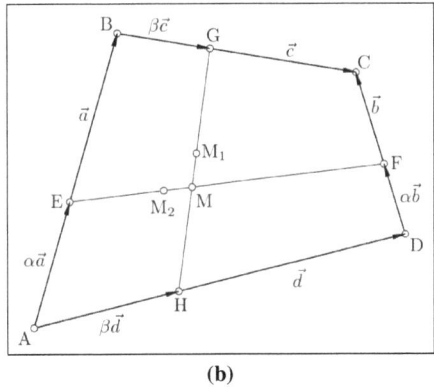

(a) (b)

Fig. 2.5: Planar convex quadrilateral

on EF and GH, so that $HM_1/M_1G = \alpha$ and $EM_2/M_2F = \beta$, $\overrightarrow{AM_1}$ and $\overrightarrow{AM_2}$ are determined from Eq. (2.16) as follows:

$$2\overrightarrow{AM_1} = \left((1+\alpha)\vec{a} + \beta\vec{c} + \beta\vec{d}\right) + \left(\frac{\alpha-1}{\alpha+1}\right)\left((1+\alpha)\vec{a} + \beta\vec{c} - \beta\vec{d}\right) \quad (2.19)$$

$$\therefore \quad 2\overrightarrow{AM_1} = \left((1+\beta)\vec{d} + (1+\alpha)\vec{b} - \vec{c} + \beta\vec{d}\right)$$
$$+ \left(\frac{\alpha-1}{\alpha+1}\right)\left((1+\beta)\vec{d} + (1+\alpha)\vec{b} - \vec{c} - \beta\vec{d}\right) \quad (2.20)$$

$$2\overrightarrow{AM_2} = \left((1+\beta)\vec{d} + \alpha\vec{b} + \alpha\vec{a}\right) + \left(\frac{\beta-1}{\beta+1}\right)\left((1+\beta)\vec{d} + \alpha\vec{b} - \alpha\vec{a}\right) \quad (2.21)$$

$$\therefore \quad 2\overrightarrow{AM_2} = \left((1+\alpha)\vec{a} + (1+\beta)\vec{c} - \vec{b} + \alpha\vec{a}\right)$$
$$+ \left(\frac{\beta-1}{\beta+1}\right)\left((1+\alpha)\vec{a} + (1+\beta)\vec{c} - \vec{b} - \alpha\vec{a}\right) \quad (2.22)$$

Subtracting the sum of Eqs. (2.19) and (2.20) from the sum of Eqs. (2.21) and (2.22) results in $\overrightarrow{AM_1} = \overrightarrow{AM_2} = \overrightarrow{AM}$. □

For grid refinement, the sides of each cell in Γ_1 are subdivided uniformly with the same number of divisions on the opposite sides. Connecting the added nodes with straight lines divides the cell into sub-cells, Fig. 2.6. The number of the subdivisions in each direction of the grid coordinates $\xi_1\xi_2$ can be different from the other. In order to maintain the grid continuity between cells, the same number of subdivisions should be applied to all cells. Theorem 2.2 assures that for a convex cell with uniform subdivision of the sides, each connecting line is subdivided uniformly and, therefore, the subdivision of a cell is a structured grid and the topology of the grid is preserved.

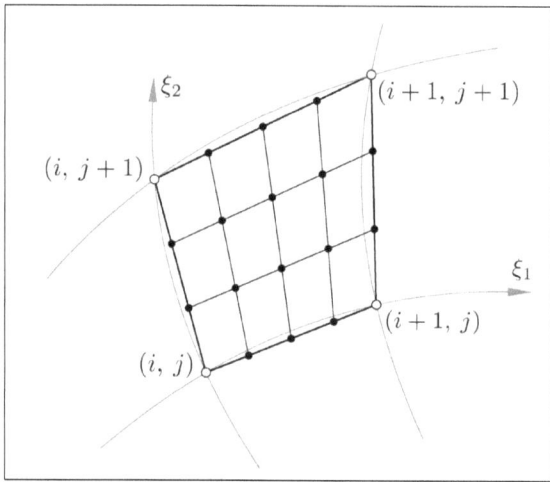

Fig. 2.6: Grid refinement
A uniform distribution of points on the opposite sides of a cell are connected with straight lines.

As shown in Fig. 2.6, the newly added nodes, marked by the black filled circles, are not necessarily on the grid lines. This refinement is used as an initialization for the back-transformation method, which adjusts all nodes to the grid coordinates. The resulting refined grid is called Γ_2 with the corresponding indexing set Γ'_2. Obviously $\Gamma_1 \subset \Gamma_2 \subset \mathbb{R}^2$.

2.2 Back-transformation

2.2.1 Introduction

The back-transformation of an image is a one-to-one transformation from the image plane to the object plane, in which each point on the object plane is assigned to a corresponding point on the image plane and vice versa. A necessary condition for this correspondence is that the image is in focus. In out of focus regions on the image plane, a point on the object plane corresponds to a region in the image plane. In this work, it is assumed that a one-to-one correspondence between the object and the image planes exists.

In some back-transformation methods, a set of points in the image plane, whose correspondence with the object plane is known, is selected and a functional fit of specific functions, such as polynomials, is used for the transformation. The selection of a correct transformation function is dependent on the type of

Chapter 2. Distortion Compensation of Digital Images

the distortion, which is generally unknown. In practice, the selection of the best transformation function is a matter of trial and error. In common optical systems, linear and second order transformations usually give satisfactory results. If applicable, back-transformation by these functions is fast and results in smooth variations. However, for distortions of higher order the use of ordinary functions is not always satisfactory.

Partial differential equations (PDEs) have found many scientific applications and are means of describing physical phenomena. Because of the analogy between certain physical phenomena, a set of PDEs may find multiple physical implementations. For instance, the Laplace equation describes certain electromagnetic, fluid flow, and heat transfer phenomena. The solution of a set of PDEs varies according to the initial and boundary conditions and can cover a large class of functions. For example, a subset of the solutions of the Navier-Stokes equations are the physical fluid phenomena, which are experienced in the daily life.

The two-dimensional Poisson equation describes the heat transfer in a plate with heat source or sink. With the lack of the heat source or sink, it reduces to the Laplace equation and in the case of the Neumann boundary conditions, the iso-temperature lines and the lines of constant heat flux will form an orthogonal grid. The introduction of a heat source or sink to the field alters the distribution of the iso-temperature and the constant heat flux lines. This property of the Poisson equation has found applications in grid generation methods. The adjustment of the source functions for grid clustering near boundaries has been used in the elliptic grid generation techniques (Steger and Sorenson, 1979). The characteristics of the Poisson equation makes it a suitable choice as the transformation function for the back-transformation of distorted images. The idea is to determine the source functions, so that the rectangular grid in the object plane be transformed into the distorted grid in the image plane.

A summary of the set definitions is given in Tab. 2.1. The grids are assumed to be structured. The indexing set of each grid is indicated by a prime ($'$), e.g. Γ'_1 for Γ_1. The coordinate axes x_1 and x_2 in the object plane are equivalent to the transformed coordinate axes ξ_1 and ξ_2 in the image plane, respectively. In each indexing set $\{(i, j)\}$, i varies along x_1 and ξ_1 axes and j along x_2 and ξ_2. It is assumed that the indexing sets of the relevant sets between the object and the image planes have equivalent coordinates. Therefore, $G'_1 \equiv \Gamma'_1$ and $G'_2 \equiv \Gamma'_2$.

The distorted grid in the image plane, Γ_1, involves the distortion information and is used to determine the source terms of the Poisson equations. The source terms are computed in G_1, whose relation with Γ_1 is known. The grid G_1 is then refined to provide a better accuracy to the transformation. The source terms are distributed on the refined grid by interpolation. The solution of the Poisson equations on the refined grid G_2 by the finite difference method (FDM) results in the transformation data, which are used for image reconstruction.

Table 2.1: Summary of the set definitions used in the image reconstruction

Grid	Description
D_1	The set of image pixels or the digital image in the image plane, Sect. 2.1.2.
D_2	The subset of D_1 with maximum local intensities, Eq. (2.7).
D_3	The set of pixels of the back-transformed image, Sect. 2.3.
I_1	The set of the light intensities assigned to the D_1 elements.
I_2	The set of the light intensities assigned to the D_3 elements.
G_1	A rectangular Cartesian grid of points in the object plane. Equivalent to Γ_1 in the image plane.
G_2	Refined G_1 with uniform subdivision of cells. Equivalent to Γ_2 in the image plane.
Γ_0	The grid of the approximate locations of the Gaussian patterns in the image plane as described in Sect. 2.1.3.
Γ_1	The distorted grid in the image plane as described in Sect. 2.1.3. Equivalent to G_1 in the object plane.
Γ_2	Refined Γ_1 as described in Sect. 2.1.4 with the same number of cell subdivisions as in G_2. Equivalent to G_2 in the object plane.

2.2.2 Mapping by the Poisson equation

Considering two subdomains D_x and D_ξ in $\mathbb{R}^2$ in the two-dimensional Euclidean space, the mapping of $(x_1, x_2) \in D_x$ into $(\xi_1, \xi_2) \in D_\xi$ by the Poisson equations is as follows (Tannehill et al., 1997):

$$\xi_i(x_j) : D_x \longmapsto D_\xi$$
$$\xi_{i,11}(x_j) + \xi_{i,22}(x_j) = \overset{i}{p}(x_j), \quad i,j \in \{1, 2\} \tag{2.23}$$

where $(x_j) \equiv (x_1, x_2)$ is a point in the coordinate system $x_1 x_2$ and $\overset{i}{p}(x_j)$ are the source functions. A comma in the subscript denotes partial differentiation:

$$\varphi_{k,mn}(x_j) := \frac{\partial^2 \varphi_k(x_j)}{\partial x_m \partial x_n}, \quad \text{for an arbitrary function } \varphi_k(x_j) \tag{2.24}$$

Equations (2.23) are a mapping from D_x into D_ξ. Generally, it transforms the Cartesian coordinates into a curvilinear coordinates. The inverse mapping can be derived as follows. Considering the inverse functions $x_i(\xi_j)$, the chain rule of differentiation results in:

$$\frac{\partial x_1}{\partial x_1} = \frac{\partial x_1}{\partial \xi_1} \frac{\partial \xi_1}{\partial x_1} + \frac{\partial x_1}{\partial \xi_2} \frac{\partial \xi_2}{\partial x_1} = x_{1,1}\, \xi_{1,1} + x_{1,2}\, \xi_{2,1} = 1 \tag{2.25}$$

In a similar way the following system of equations is achieved:

$$\begin{cases} x_{1,1}\,\xi_{1,1} + x_{1,2}\,\xi_{2,1} = 1 \\ x_{1,1}\,\xi_{1,2} + x_{1,2}\,\xi_{2,2} = 0 \\ x_{2,1}\,\xi_{1,1} + x_{2,2}\,\xi_{2,1} = 0 \\ x_{2,1}\,\xi_{1,2} + x_{2,2}\,\xi_{2,2} = 1 \end{cases} \tag{2.26}$$

Solution for $\xi_{i,j}$ gives:

$$\xi_{1,1} = \;\;\; x_{2,2}/\tilde{J} \tag{2.27}$$
$$\xi_{1,2} = -x_{1,2}/\tilde{J} \tag{2.28}$$
$$\xi_{2,1} = -x_{2,1}/\tilde{J} \tag{2.29}$$
$$\xi_{2,2} = \;\;\; x_{1,1}/\tilde{J} \tag{2.30}$$

where $\tilde{J}$ is the inverse Jacobian of the transformation:

$$\tilde{J} = x_{1,1} x_{2,2} - x_{1,2} x_{2,1} \tag{2.31}$$

From (2.25):

$$\begin{aligned}\frac{\partial^2 x_1}{\partial x_1^2} &= x_{1,11}\,\xi_{1,1}^2 + 2\,x_{1,12}\,\xi_{1,1}\,\xi_{2,1} + x_{1,22}\,\xi_{2,1}^2 \\ &\quad + x_{1,1}\,\xi_{1,11} + x_{1,2}\,\xi_{2,11} = 0\end{aligned} \tag{2.32}$$

Similarly:

$$\begin{aligned}\frac{\partial^2 x_1}{\partial x_2^2} &= x_{1,11}\,\xi_{1,2}^2 + 2\,x_{1,12}\,\xi_{1,2}\,\xi_{2,2} + x_{1,22}\,\xi_{2,2}^2 \\ &\quad + x_{1,1}\,\xi_{1,22} + x_{1,2}\,\xi_{2,22} = 0\end{aligned} \tag{2.33}$$

Adding both sides of these equations and inserting from Eq. (2.23) gives:

$$\begin{aligned} & x_{1,11}\,(\xi_{1,1}^2 + \xi_{1,2}^2) + 2\,x_{1,12}\,(\xi_{1,1}\,\xi_{2,1} + \xi_{1,2}\,\xi_{2,2}) + x_{1,22}\,(\xi_{2,1}^2 + \xi_{2,2}^2) \\ & \quad + x_{1,1}\,\overset{1}{p} + x_{1,2}\,\overset{2}{p} = 0 \end{aligned} \tag{2.34}$$

Inserting Eqs. (2.27) to (2.31) in Eq. (2.34) gives the final form of inverse mapping as:

$$x_i(\xi_j) : \mathbf{D}_\xi \longmapsto \mathbf{D}_x$$
$$\alpha\, x_{i,11} - 2\,\beta\, x_{i,12} + \gamma\, x_{i,22} = -\tilde{J}^2\, x_{i,k}\, \overset{k}{q}, \quad i,\,j,\,k \in \{1,\,2\} \tag{2.35}$$

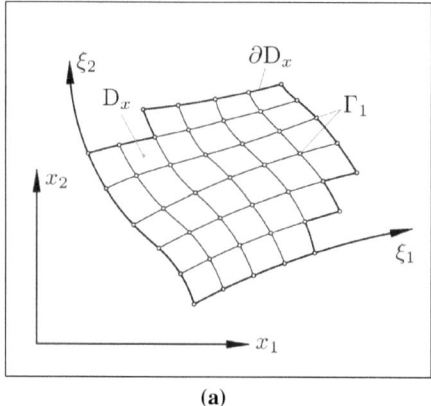 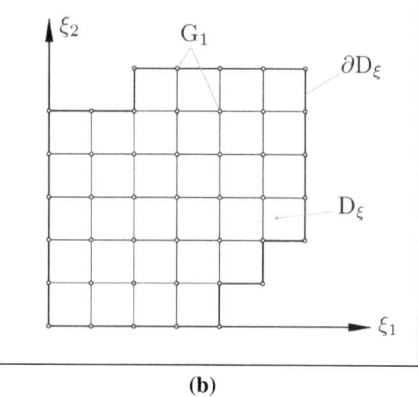

Fig. 2.7: Back-transformation (a) domain D_x containing the distorted calibration grid Γ_1 in the image plane and (b) the transformed domain D_ξ
The goal of back transformation is to map Γ_1 into a rectangular grid G_1.

$$\alpha(\xi_j) := x_{1,2}^2 + x_{2,2}^2 \tag{2.36}$$

$$\beta(\xi_j) := x_{1,1}\, x_{1,2} + x_{2,1}\, x_{2,2} \tag{2.37}$$

$$\gamma(\xi_j) := x_{1,1}^2 + x_{2,1}^2 \tag{2.38}$$

where $\overset{i}{q}(\xi_j) := \overset{i}{p}(x_k(\xi_j))$ and the Einstein's summation convention[1] is valid for the repeated indices. If the source functions $\overset{i}{q}$ vanish identically to zero, the Poisson equations reduce to the Laplace equations.

Figure 2.7 shows an image plane with a distorted calibration grid Γ_1 within a domain D_x. The goal of back-transformation is to find a transformation, so that Γ_1 be mapped onto the rectangular grid of a calibration image G_1. For proper source functions $\overset{i}{q}$, the solution of the inverse Poisson equations (2.35) in D_ξ results in the mapping data.

In the following section, a numerical solution of the inverse Poisson equations using the finite difference method (FDM) is presented. Since D_ξ is not, in general, a rectangular domain, an explicit algorithm is used to solve the set of the discretized Poisson equations.

---

[1] A repeated index in a term which appears once as an upper and once as a lower index implies summation over the whole range of the index.

Chapter 2. Distortion Compensation of Digital Images

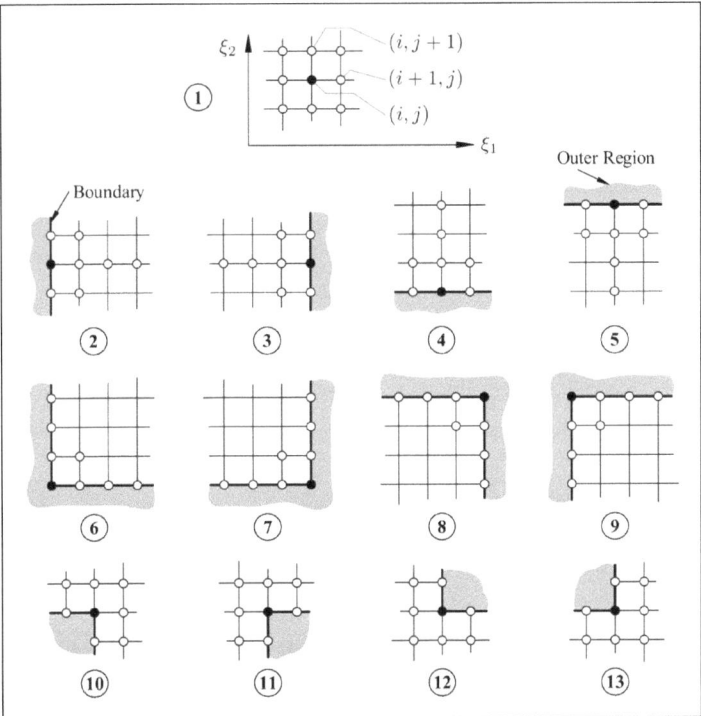

Fig. 2.8: Discretization patterns for the rectangular grid of domain D_ξ
Pattern 1: an internal node with coordinates and indices and **Patterns 2–13:** the boundary nodes. The black circles show the principle nodes and the hollow circles show the auxiliary nodes, which are used in each discretization.

2.2.3 Discretization

Figure 2.8 shows different possible patterns for a finite difference discretization. The patterns stand for second order discretization at the internal nodes and second or mixed first and second order discretizations at the boundary nodes. A summary of the relevant discrete equations is given in App. E.

At an internal node (i, j), pattern 1 in Fig. 2.8, the discrete form of Eq. (2.35) is as follows:

$$\frac{\alpha|_{i,j}}{\Delta\xi_1^2} \left(x_m|_{i+1,j} - 2\, x_m|_{i,j} + x_m|_{i-1,j} \right)$$

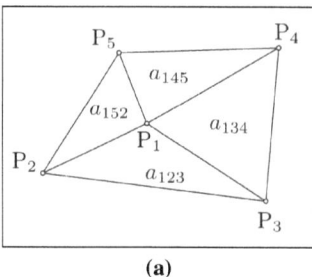 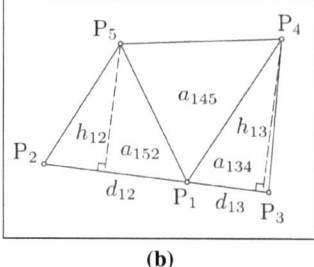

Fig. 2.9: Interpolation

$$-\frac{\beta|_{i,j}}{2\,\Delta\xi_1\,\Delta\xi_2}\left(x_m|_{i+1,j+1} - x_m|_{i+1,j-1} - x_m|_{i-1,j+1} + x_m|_{i-1,j-1}\right)$$
$$+\frac{\gamma|_{i,j}}{\Delta\xi_2^2}\left(x_m|_{i,j+1} - 2\,x_m|_{i,j} + x_m|_{i,j-1}\right)$$
$$= -\bar{J}^2\Big|_{i,j}\,x_{m,k}|_{i,j}\,\overset{k}{q}\Big|_{i,j}\,, \qquad m,\,k \in \{1,\,2\} \tag{2.39}$$

in which $\Delta\xi_1$ and $\Delta\xi_2$ are grid spacings in the ξ_1 and ξ_2 directions, respectively. For a typical discrete variable φ, $\varphi|_{i,j}^n$ is its value at node $(i,\,j)$ and at time step or iteration step n, which may be omitted as in the above equations. The discretizations at the boundary nodes are derived in the same manner, resulting in different variants of Eq. (2.39).

2.2.4 Computation of the source functions

Since the physical grid G_1 in the object plane as well as the distorted grid Γ_1 in the image plane are known, the source terms $\overset{1}{q}$ and $\overset{2}{q}$ can be determined by solving the discretized form of Eq. (2.35) in G_1.

The distribution of the source functions inside G_2 is determined from that in G_1 by interpolation. For a given cell $P_2P_3P_4P_5$ as a convex quadrilateral encompassing a grid node P_1, Fig. 2.9a, the value of a function $\varphi(x_1,\,x_2)$ at P_1, denoted by φ_1, is computed by interpolating its values at the cell vertices (Mohseni, 2000):

$$\varphi_1 = \frac{a_{134}a_{145}\varphi_2 + a_{145}a_{152}\varphi_3 + a_{152}a_{123}\varphi_4 + a_{123}a_{134}\varphi_5}{a_{134}a_{145} + a_{145}a_{152} + a_{152}a_{123} + a_{123}a_{134}} \tag{2.40}$$

where a_{1mn} is the area of the triangle $\triangle P_1P_mP_n$:

$$a_{1mn} = \frac{1}{2}\left|\overrightarrow{P_1P_m} \times \overrightarrow{P_1P_n}\right| \tag{2.41}$$

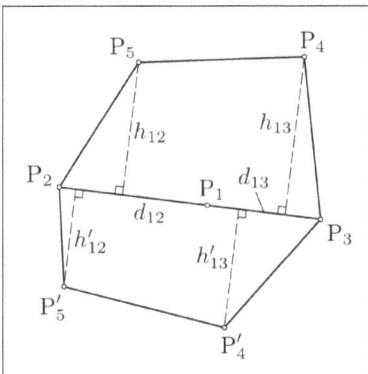

Fig. 2.10: Interpolation along the common side of two adjacent cells d_{12} and d_{13} are the lengths of the segments P_1P_2 and P_1P_3, respectively.

The interpolation preserves the vertex values. On the cell sides, Eq. (2.40) turns into the interpolation of the function values at the end vertices weighted with the distances of other vertices from the side. For example on P_2P_3, Fig. 2.9b:

$$\varphi_1 = \frac{d_{13}\, h_{13}\, \varphi_2 + d_{12}\, h_{12}\, \varphi_3}{d_{13}\, h_{13} + d_{12}\, h_{12}} \tag{2.42}$$

where d_{12} and d_{13} are the lengths of P_1P_2 and P_1P_3, and h_{12} and h_{13} are the vertical distances of P_5 and P_4 from P_2P_3, respectively.

In general, the continuity of the interpolation on the cell sides is not preserved except at the cell vertices. The conditions required for the continuity of the interpolated function across the cell boundaries is provided by the following theorem:

Theorem 2.3 Let $P_2P_3P_4P_5$ and $P_2P_3P'_4P'_5$ be two adjacent cells as shown in Fig. 2.10 and φ a function to be interpolated at P_1, whose values at the cell vertices are given as φ_2, φ_3, ..., at P_2, P_3, ..., respectively. Then the interpolated values of φ_1 in each cell are equal along P_2P_3 if and only if $\varphi_2 = \varphi_3$ or $h_{12}/h_{13} = h'_{12}/h'_{13}$. □

Since the interpolation is continuous, smooth, and preserves the vertex values, the difference between the interpolated values on the cell sides is finite and is dependent on the cell distortion. The discontinuity would be negligible for slightly distorted cells. In highly distorted regions of an image, it might be required to use a more sophisticated interpolation algorithm, which maintains the continuity of the interpolated function across the cells. However, an increase of the number of the grid points in the original calibration grid in the object plane, G_1, can moderate the discontinuity effect on the cell sides.

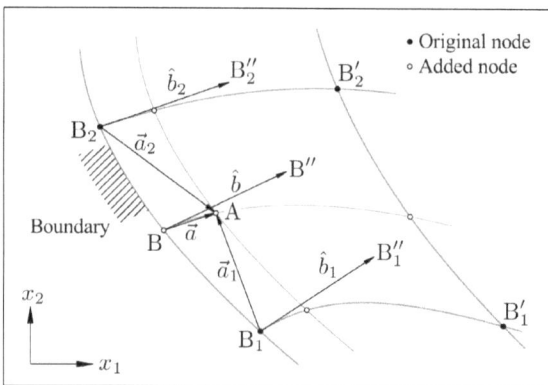

Fig. 2.11: Application of the boundary conditions
On the boundary curve B_1BB_2, B_1 and B_2 are the original nodes from Γ_1 and are fixed. The location of an added node B is determined from the location of its neighbor node A by the interpolated branch slope on the boundary curve.

2.2.5 Boundary and internal conditions

Boundary conditions are applied to the boundary nodes in the image plane. The Dirichlet boundary condition is applied to the original grid points of Γ_1 in Γ_2, with leaving the location of the nodes fixed.

The added nodes on the boundaries are adjusted by the branch slopes. Figure 2.11 shows a boundary region in the image plane. The nodes B_1, B_2, B_1', and B_2' are original nodes from Γ_1. The boundary curve B_1BB_2 and the slopes of the branches from the original nodes on it, i.e. the slopes of the unit vectors $\hat{b}_1$ and $\hat{b}_2$, are approximated by a natural cubic spline fit (for details see App. B). An added node B on the boundary near an internal node A is located so that the vectors $\vec{a}$ and $\hat{b}$ be collinear. Here $\hat{b}$ is a unit vector at B, whose slope is equal to the interpolated slope along the boundary.

During the solution, the original grid Γ_1 as well as the original nodes at the boundaries in Γ_2, which are subject to the Dirichlet boundary condition, remain fixed. The locations of the internal nodes as well as the added nodes on the boundaries in Γ_2 are modified during the solution. In each iteration, the distribution of source the functions in Γ_2 is computed from the original distribution in Γ_1 by interpolation.

2.2.6 Solution of the Poisson equations

Assuming that G_2 is a refinement of G_1 by adding N_1 nodes in the ξ_1 direction and N_2 nodes in the ξ_2 direction in each cell, for given source functions $\overset{1}{q}|_{i,j}$ and $\overset{2}{q}|_{i,j}$ in G_2 the grid coordinates x_m can be computed from Eq. (2.39) at an internal node (i, j) iteratively using the following explicit equation:

$$
\begin{aligned}
x_m|_{i,j}^{n+1} = \Bigg[& \frac{N_1'^2 \, \alpha|_{i,j}^n}{\Delta \xi_1^2} \left(x_m|_{i+1,j}^n + x_m|_{i-1,j}^n \right) \\
& - \frac{N_1' N_2' \, \beta|_{i,j}^n}{2 \, \Delta \xi_1 \, \Delta \xi_2} \left(x_m|_{i+1,j+1}^n - x_m|_{i+1,j-1}^n \right. \\
& \left. \quad - x_m|_{i-1,j+1}^n + x_m|_{i-1,j-1}^n \right) \\
& + \frac{N_2'^2 \, \gamma|_{i,j}^n}{\Delta \xi_2^2} \left(x_m|_{i,j+1}^n + x_m|_{i,j-1}^n \right) \\
& + \tilde{J}^2 \Big|_{i,j}^n \; x_{m,k}|_{i,j}^n \; q\Big|_{i,j}^{k\,n} \Bigg] \\
\Bigg/ & \left(\frac{2 N_1'^2 \, \alpha|_{i,j}^n}{\Delta \xi_1^2} + \frac{2 N_2'^2 \, \gamma|_{i,j}^n}{\Delta \xi_2^2} \right), \qquad m, k \in \{1, 2\}
\end{aligned} \qquad (2.43)
$$

where for $m \in \{1, 2\}$, $\Delta \xi_m$ is the grid spacing in G_1, $N_m' := N_m + 1$, and $n \in \mathbb{N}$ is the step of iteration. After initialization, the solution procedure consists of an iteration loop as in the following pseudo-code:

```
repeat:
    Solve Eq. (2.43) at all nodes.
    Apply the boundary conditions.
    Update the source functions at the added nodes.
    Update the grid data structure.
until the convergence criterion is passed.
```

2.3 Image reconstruction

The solution of the transformation equations results in a refined grid Γ_2 as a transformation of G_2. The transformation is applicable to the PIV images with the same setup as the calibration image. During the image reconstruction, the rectangular image pixels are generally transformed into non-rectangular regions, Fig. 2.12. Therefore, in order to improve the transformation quality, it is required that the distorted image be subdivided to achieve higher resolution. To this end, a reconstructed image D_3 is defined on D_1, Eq. (2.4), with an adjustable resolution.

Using a search algorithm and interpolation, the location of each pixel in D_3 is computed. This procedure consists of finding a cell in Γ_2, which contains a point in D_3. The coordinates of a pixel in the image plane, Γ_2, is approximated by the interpolation Eq. (2.40). The intensity value at each found location can be directly assigned to the pixel. By repeating the same procedure for all pixels in D_3, the reconstructed image is gained.

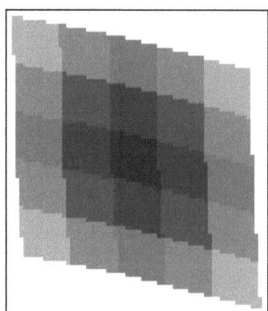

Fig. 2.12: Deformation of pixels after reconstruction
Each deformed pixel is represented by the smaller pixels, whose sizes are identifiable from the saw-toothed patterns on the boundaries.

2.4 Applications

The applications of the transformation method are presented in this section. Some characteristics of the transformation are presented by examples. The gray scales of the images are inverted for printing purposes.

Figure 2.13 shows a synthetic grid generated by the equidistant translations of two smooth curves along straight lines. The grid contains nearly rectangular cells as well as cells with skewness to the right (right) and to the left (top left). The grid nodes are marked by the symmetric Gaussian distributions, whose extrema coincide with the node locations. The Gaussian patterns are the same for all nodes. The grid has a relative poor density in the regions of high grid line curvature (middle right and middle top).

Figure 2.14a shows the identified original grid (black) and the refined grid (gray) with four added nodes in each direction. The refined grid is the result of the solution of the Poisson equations and coincides with the original grid only at the boundary points. The adjustment of its internal nodes is dependent on the source functions, which allow the arrangement of the added nodes along curves rather than lines and a better approximation to the original grid. After the identification of the original grid, its boundaries are approximated by the natural cubic splines. The locations of the added nodes on the boundaries are confined to these splines during the application of the boundary conditions as explained in Sect. 2.2.5.

The distribution of the distances between the original nodes and their corresponding nodes in the refined grid, $\delta_{i,j}$, after the solution of the Poisson equations are shown on the original grid in Fig. 2.14b. For a node $(x_1|_{i,j}, x_2|_{i,j}) \in \Gamma_1$ and its corresponding node in the refined grid $(\tilde{x}_1|_{m,n}, \tilde{x}_2|_{m,n}) \in \Gamma_2$, $\delta_{i,j}$ is calculated as follows:

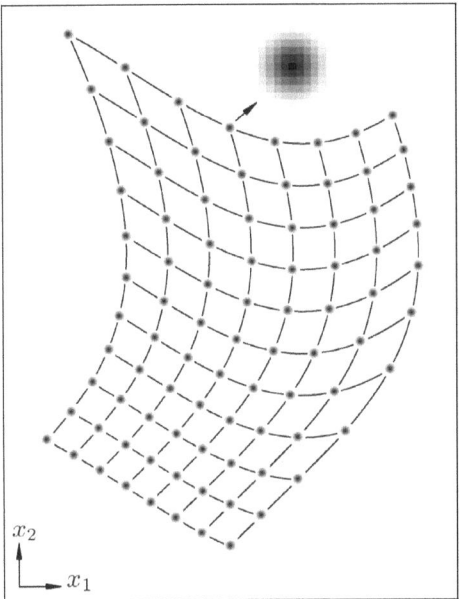

Fig. 2.13: Synthetic grid from smooth curves
The grid is generated by the equidistant translations of two crossing curves along straight lines. The grid nodes are marked by two-dimensional Gaussian patterns, whose maxima coincide with the node locations.

$$\delta_{i,j} := \sqrt{\left(x_1|_{i,j} - \tilde{x}_1|_{m,n}\right)^2 + \left(x_2|_{i,j} - \tilde{x}_2|_{m,n}\right)^2} \in \mathbb{R}^\oplus \tag{2.44}$$

Since the original nodes at the boundaries are subject to the Dirichlet boundary conditions, the differences between the original nodes and their corresponding nodes in the refined grid at the boundaries are zero. In a considerable region, the distance remains under one pixel, while in the regions of high grid line curvature it increases up to 2.4 pixels. This is a measure of the accuracy of the transformation in the range of 600 pixels in x_1 and 800 pixels in x_2 directions for a 8×11 nodes grid.

The distribution of the source terms in the image and the object planes are shown in Figs. 2.15 and 2.16, respectively. The source terms are computed on the original grid and then are interpolated on the added nodes in the refined grid. Along the boundaries, the source terms at the added nodes do not participate in the solution and have been optionally set to zero. The transformed coordinate lines in the object plane are shown in Fig. 2.17a.

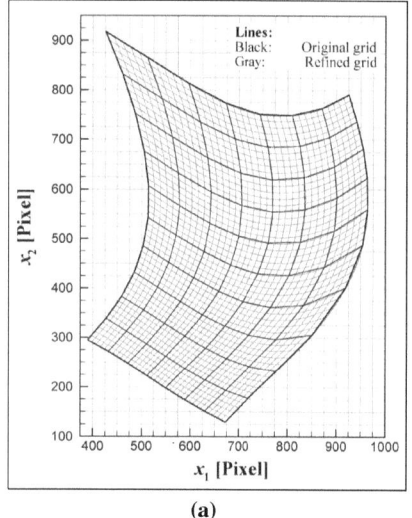 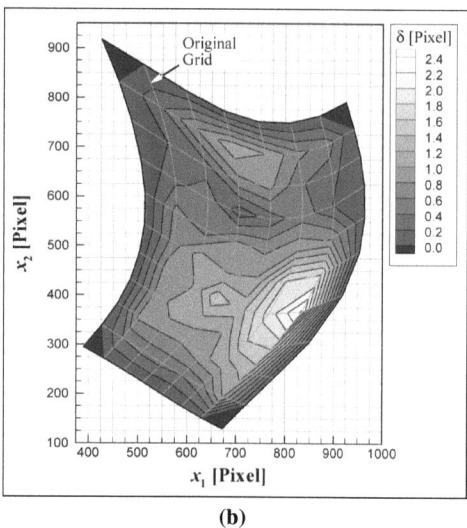

Fig. 2.14: (a) the original grid (black) and the refined grid (gray) after grid identification and the solution of the Poisson equations on the grid of Fig. 2.13 and (b) the distribution of the distances between the nodes in the original grid and their corresponding nodes in the refined grid

The reconstructed image of the grid in Fig. 2.13 is shown in Fig. 2.17b. It shows the ability of the method to resolve high distortions with a relative low grid density. Departures of the grid lines from the straight lines are visible as local distortions.

In the next example, the sensitivity of the method to nonuniform grids with low grid density is studied. Figure 2.18 shows a synthetic grid generated by the nonuniform displacements and deformations of two crossing smooth base curves. Each curve is slightly deformed in a different way compared with its neighbors. Missing nodes at the top right add extra corners to the boundaries. The density of the nodes along each grid line is so that extra nodes are needed to estimate their distribution. The grid nodes are marked by the Gaussian patterns as in Fig. 2.13

Figure 2.19a shows the original (black) and the refined (gray) grids. The deviation of the refined grid from the original grid is shown in Fig. 2.19b. They represent the effect of the grid resolution and the curvature of the grid lines on the accuracy of the transformation. Since the method uses a discretization on the original grid, the lack of nodes at the upper right corner does not limit its applicability. Due to the non-uniformity and the low resolution of the original grid, the deviation of the refined grid from the original is considerable. The maximum deviation in the 8×8 nodes grid and in the range of 600

Chapter 2. Distortion Compensation of Digital Images

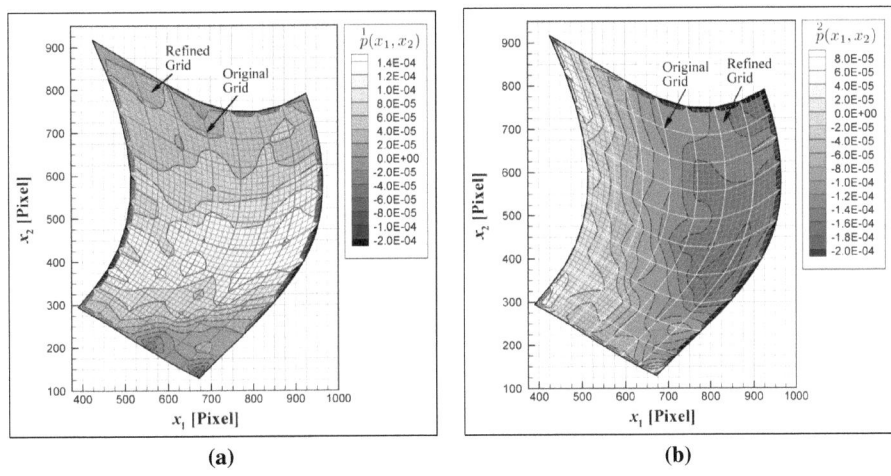

Fig. 2.15: The distribution of the source functions in the image plane

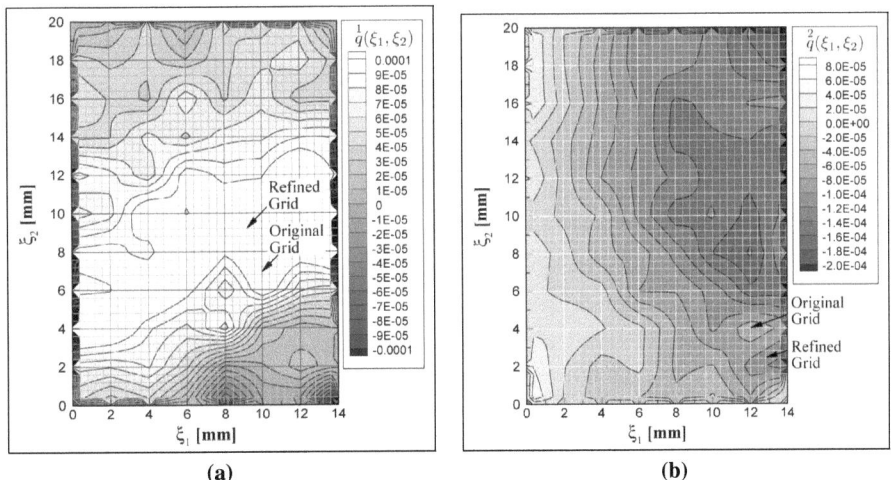

Fig. 2.16: The distribution of the source functions in the object plane

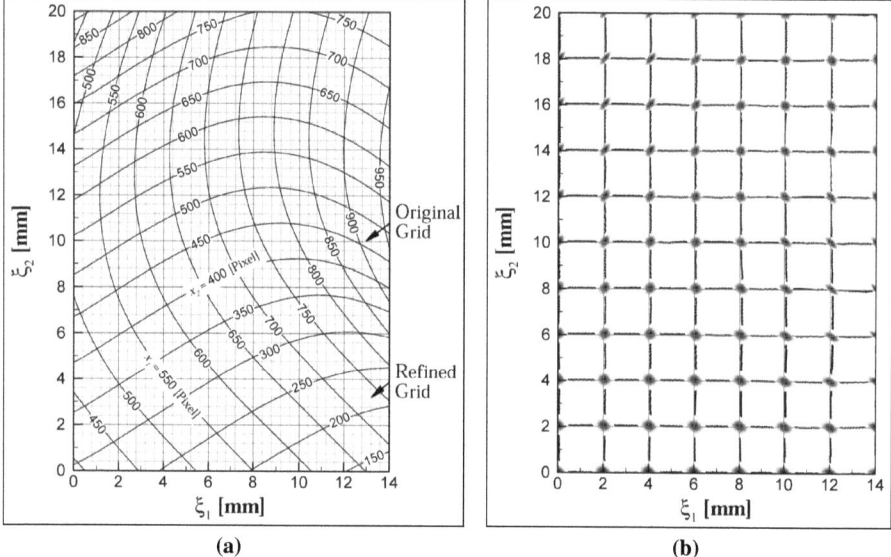

Fig. 2.17: (a) The coordinate lines after transformation and (b) the reconstructed grid
The grid lines show the expected locations of the point patterns.

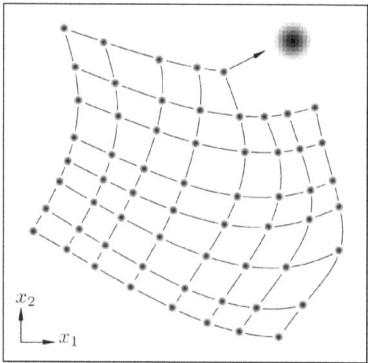

Fig. 2.18: Synthetic grid from smooth curves
The grid is generated by the nonuniform displacements and deformations of two crossing base curves. Each curve is slightly deformed in comparison with its neighbors. The grid nodes are marked by two-dimensional Gaussian patterns, whose maxima coincide with the node positions.

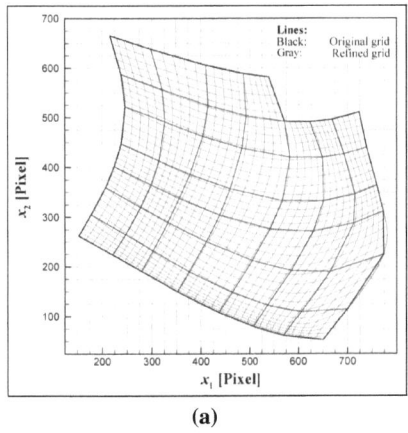 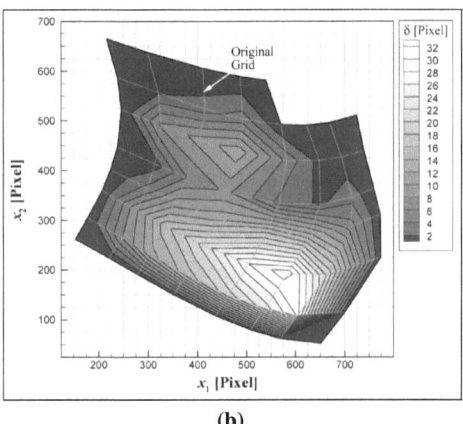

Fig. 2.19: (a) the original grid (black) and the refined grid (gray) after grid identification and the solution of the Poisson equations on the grid of Fig. 2.18 and (b) the distribution of the distances between the nodes in the original grid and their corresponding nodes in the refined grid

pixels × 600 pixels is 32 pixels, Fig. 2.19b.

The distribution of the source functions in the image plane is shown in Fig. 2.20. Figure 2.21 represents the reconstructed grid. This example shows that although the distorted grid, Fig. 2.18, does not contain enough information for the back-transformation, the method is capable of an approximate reconstruction.

The next example, Fig. 2.22, is a synthetic grid made by twisting a rectangular grid around its center point. The center of the twist is a singular point. The grid lines are smooth and identifiable from the distribution of the nodes. Despite the previous examples, the grid markings are not Gaussian patterns and are partly saturated. The saturated regions of marking patterns do not provide information about the location of the nodes. Therefore, the node coordinates are determined to within the size of the saturated regions. This example shows the effect of the quality of node markings and the ability of the method to transform a region around a singularity.

The results of the grid identification and the solution of the transformation equations are presented in Fig. 2.23. The accuracy of the transformation is within six pixels and the highest error is in the vicinity of the singularity. Considering the distribution of the source terms, Figs. 2.24 and 2.25, the increase of the resolution of the original grid or the use of a higher order interpolation of the source terms at the added nodes can improve the accuracy of the transformation. The distribution of the coordinate lines

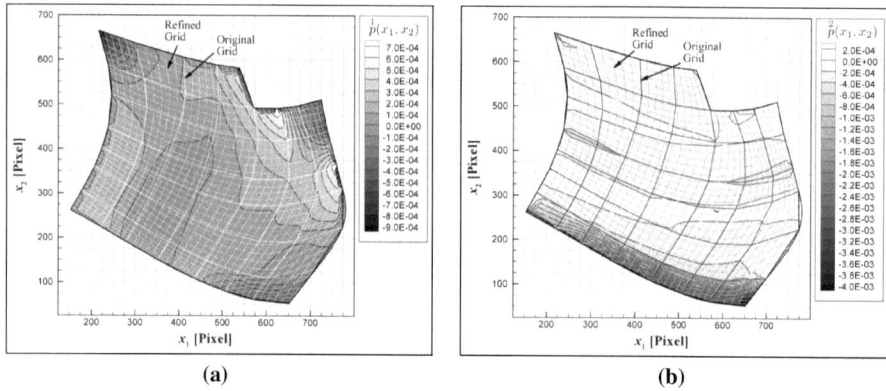

Fig. 2.20: The distribution of the source functions in the image plane

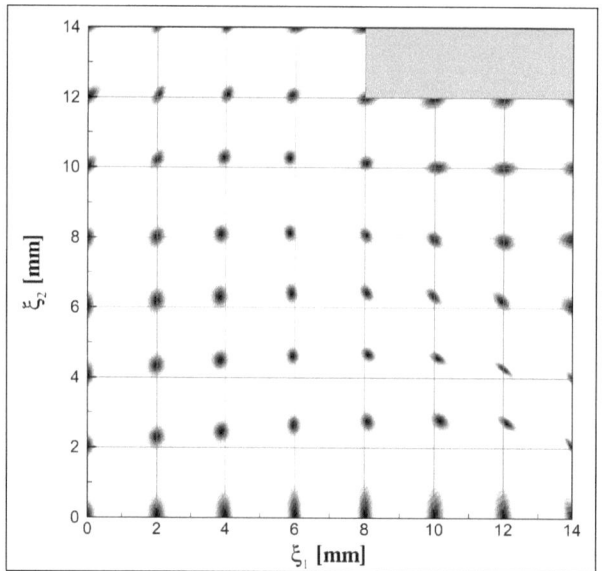

Fig. 2.21: Reconstructed grid
The grid lines show the expected locations of the point patterns.

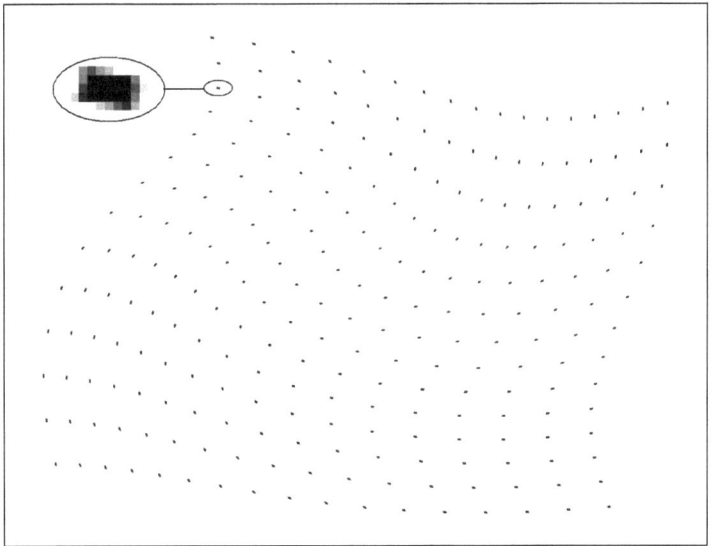

Fig. 2.22: A synthetic grid, generated by uniform distortion of a rectangular grid
The grid nodes are marked by patterns with partly saturated pixels instead of Gaussian patterns. The saturation of pixels reduces the accuracy of the determination of the node coordinates, which is a source of error in back-transformation.

and the reconstructed grid are shown in Fig. 2.26.

In this example the grid and the image characteristics of the original grid are comparable to that of the first example, Fig. 2.13. Hence, the expected mismatch between the original and the refined grids should be less than three pixels. The maximum deviation of six pixels in Fig. 2.23b is within the size of the saturated regions of the node markings and is located in the region near the singularity, where the resolution of the grid is more effective in the accuracy of the transformation. Besides the grid resolution, the saturation of the node markings affects the mismatch between the original and the refined grid and is a source of error in the transformation. The example also shows the applicability of the transformation method to such distortions with local node dislocations.

Depending on the type of distortion, the size of image, and the applied grid refinement, the iterative solution of the finite difference discretization of the Poisson equations (2.43) may take considerable time. In this regard, the distortion compensation method does not suit real-time data manipulation. However, once the transformation data are available, the image reconstruction phase can be implemented in real-time data processing.

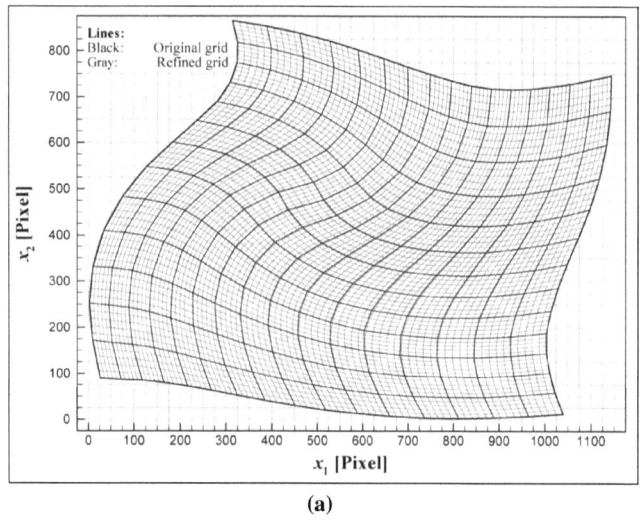

(a)

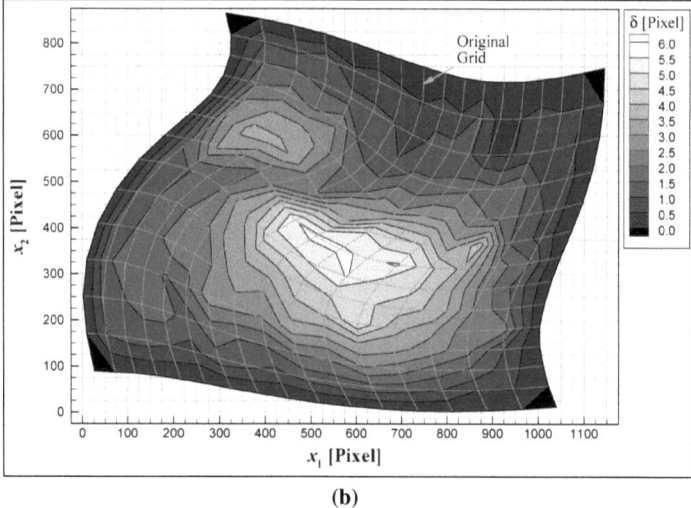

(b)

Fig. 2.23: (a) the original grid (black) and the refined grid (gray) after grid identification and the solution of the Poisson equations on the grid of Fig. 2.22 and (b) the distribution of the distances between the nodes in the original grid and their corresponding nodes in the refined grid

Chapter 2. Distortion Compensation of Digital Images

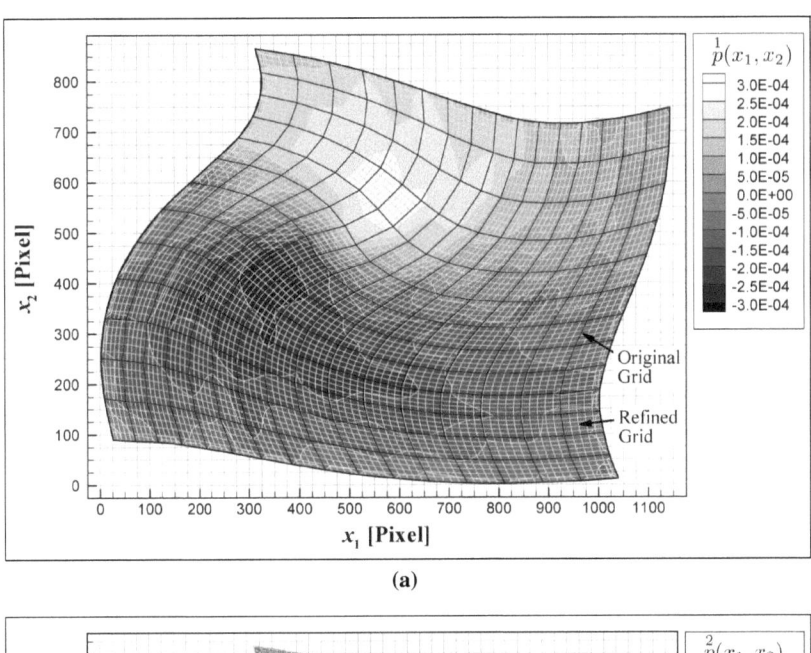

(a)

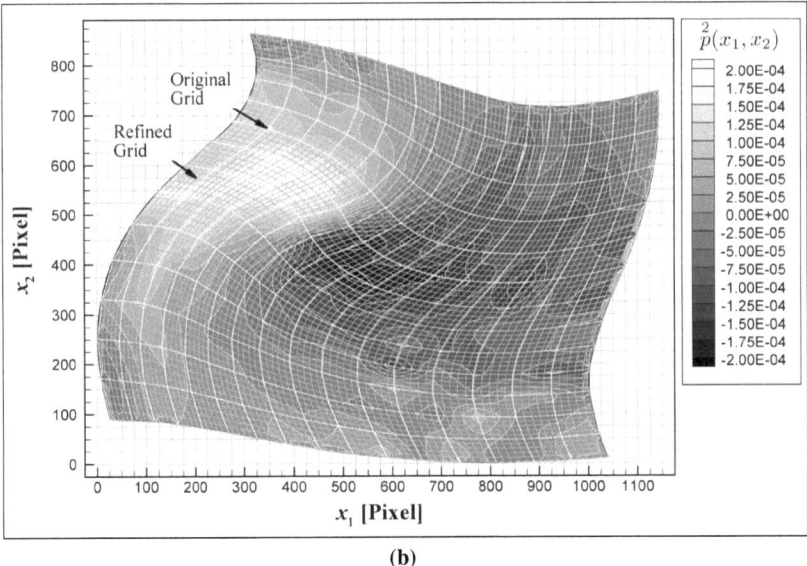

(b)

Fig. 2.24: The distribution of the source functions in the image plane

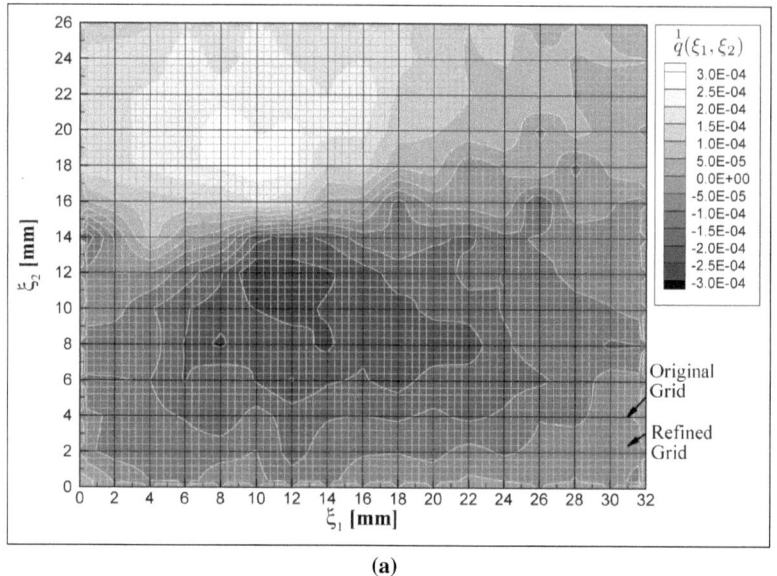

(a)

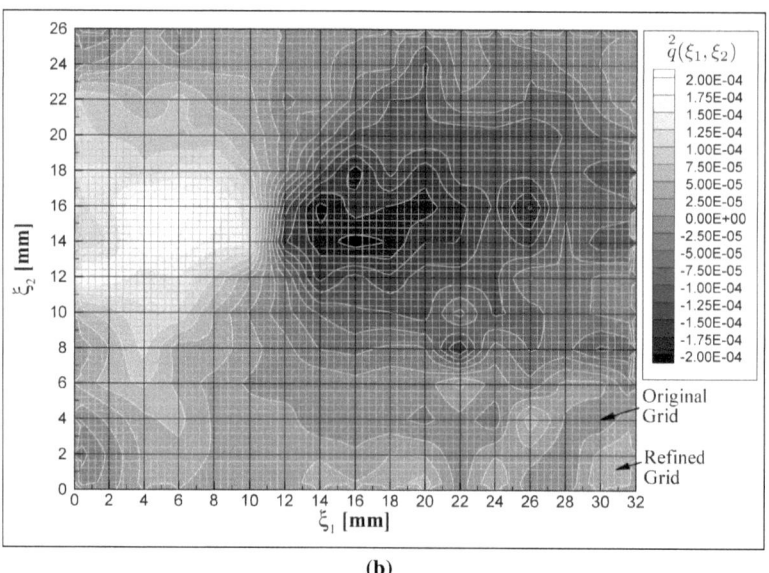

(b)

Fig. 2.25: The distribution of the source functions after transformation

The time of distortion compensation phase can be optimized in several ways. One method is the implementation of other solution algorithms instead of the explicit algorithm used. Since the solution domain is not rectangular, an explicit method is the first choice. Hybrid explicit and implicit methods may be implemented, in order to improve the convergence speed. Besides the use of acceleration and damping factors in the solution code, an estimation of the required accuracy and the optimization of the convergence criterion based on the behavior of the convergence curve can considerably affect the time of computations.

Another characteristic that affects the time and accuracy of the solution is the presence of local dislocations at the grid points. This may happen if the quality of the node markings is not good enough to estimate the locations of the nodes correctly, such as in Fig. 2.22. From the physical realizations, it is known that the Poisson equation governs smooth variations. Therefore, the presence of local node dislocations increases the tendency of the solution to diverge and the time of convergence. As shown for the distorted image of Fig. 2.18, the method is capable of adapting itself to local structures in a distorted grid. The adaptation to local dislocations is, however, a departure from the real distortion function of the imaging system. Therefore, the accuracy of the identification of the node locations affects the accuracy and the computation time of the distortion compensation method.

In this section the application of the transformation method to three synthetic grids with different characteristics was presented. The application of the method to the distorted endoscopic SPIV images is presented in Sect. 5.2.

2.5 Conclusions

The case studies in the previous section and the application of the method for the reconstruction of the calibration images of the endoscopic SPIV setup in Ch. 5, show the capability of the method to reconstruct highly distorted images. After the identification of the grid in a calibration image, the method provides a unique solution without the need for the user interference or judgment.

As a typical characteristic of iterative numerical methods, the solution algorithm takes considerably more computational time than similar analytic reconstruction methods and, therefore, is not suitable for real-time transformations. However, once the transformation data are available for a calibration configuration, the image reconstruction can be implemented in on-line data processing algorithms.

Partial differential equations as transformation functions provide a new class of image reconstruction methods, which can improve the data analysis in PIV and other image-based optical measurement techniques.

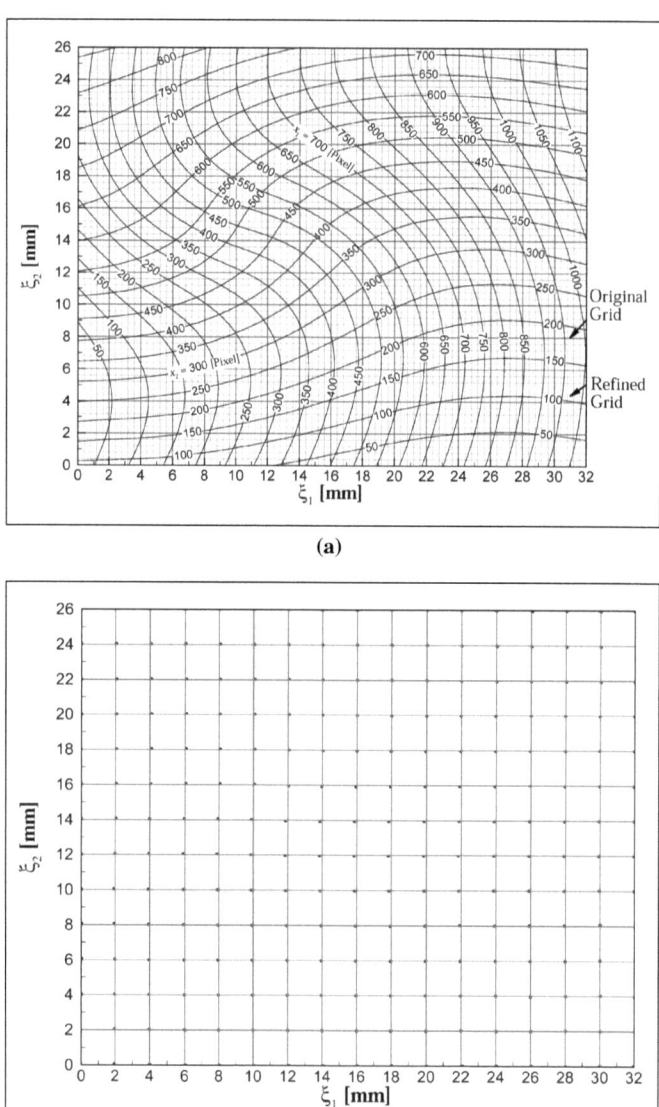

Fig. 2.26: (a) The coordinate lines after transformation and (b) the reconstructed grid. The grid lines show the expected locations of the point patterns.

Chapter 3

Theory of Local Measurements under Quasi-Steady Conditions

3.1 Introduction

With the increase of the resolution of the measurement devices, their sensitivity to the external effects, such as the ambient noise, increases. Especially during calibration, the achievement of stable conditions becomes troublesome. Even slight changes in the effective environmental conditions may be revealed in the measured data.

This problem can be dealt with in several ways. The basic approach would be to isolate the measuring system to get a stable signal, sensitive only to the variations of the measured quantities. This method is, however, not always practical, especially for an on-site calibration.

In most engineering applications, minor changes due to the environmental effects are negligible in regard to the expected accuracy of measurements. In regular calibration procedures, it is tried to stabilize a signal and to determine its statistical properties based on direct averaging. In this case, slight changes in the signal mean may cause unreal extension of its range of uncertainty. However, they may be neglected, if they are within the range of an expected accuracy. In this way at the cost of lowering the accuracy, the calibration procedure may become easier, faster, and more versatile.

In steady state measurements, only low-frequency changes of a measured signal are important and the output of the measurement system is a low-pass filtered signal. This characteristic raises the idea of the quasi-steady calibration, in which minor external and environmental effects are considered so that the accuracy of calibration is increased.

The theory presented in this chapter provides the means for the calibration of the measurement loops under slightly varying or quasi-steady conditions. In practice the achievement of steady state conditions may take long time or may not be achieved due to the presence of external or internal disturbances, such as slight changes in the ambient conditions. By including these changes in a calibration procedure, the method provides the statistical characteristics of a signal, while keeping its uncertainty range to that of the signal noise and the device accuracy.

In the following the mathematical basis for the calibration under quasi-steady conditions, including the statistical analysis and the error estimation, is presented. The application of the method to the flow measurements by the aerodynamic pressure probes is presented in the next chapter.

3.2 Measurement

Under steady state conditions, field variables like pressure or temperature are time independent. In practice, such as flow measurements in turbomachinery, steady state conditions are hardly achieved. Under most stable conditions, the field variables of a flow are approximately statistically stationary and are weakly time dependent in the mean.

If the rate of change of a field variable compared to the time constants[1] of a measurement system is negligible, steady state measurements can still be performed. Generally, a measured signal, acquired at a constant location, can be decomposed into a mean variation and a superposed fluctuation or noise. In an ideal steady state, the signal is identical to its mean and is time independent. In practice, however, noise usually accompanies a signal and under stable conditions, only the mean of the signal is time independent.

A stable system, i.e. a system which is in some equilibrium state, is usually sensitive to external influences. For such a system, steady conditions can hardly be perfectly achieved in most measurements. If the deviation from steady state is slow compared to the time constants of the system (including the measurement devices), the measurements can be considered quasi-steady, meaning that field variables are weakly time dependent in the mean and may contain noise.

In the following, a mathematical approach for the data analysis of quasi-steady measurements is presented. A field variable is first decomposed into a mean and a fluctuation or noise. Quasi-steady condition is then defined for slight deviations from the steady state and its statistical analysis is presented. In this work, the noise of a signal is assumed to be stochastic in nature and tend to vanish in the mean.

Definition 3.1 The *special time average* of a continuous field variable $\varphi(\boldsymbol{x}, t)$ at a point $\boldsymbol{x} \in \mathbb{R}^3$ and at a time t is defined as:

$$\bar{\varphi}(\boldsymbol{x}, t) := \frac{1}{t_2' - t_1'} \int_{t+t_1'}^{t+t_2'} \varphi(\boldsymbol{x}, t') \, dt' , \quad t + t_1' \leq t \leq t + t_2' \tag{3.1}$$

---

[1] The *time constant* of a field or a device variable is the required time after a change or disturbance for the variable to become stable within a defined neighborhood of its steady state.

where $[t'_1, t'_2]$ is an arbitrary time span selected for averaging. □

With this definition, a field variable $\varphi(\boldsymbol{x}, t)$ can be decomposed into a time average and a fluctuation:

$$\varphi(\boldsymbol{x}, t) = \bar{\varphi}(\boldsymbol{x}, t) + \bar{\varphi}'(\boldsymbol{x}, t) \qquad (3.2)$$

$\bar{\varphi}'(\boldsymbol{x}, t)$ is called the *noise* of the field variable.

The selection of the averaging time span $[t'_1, t'_2]$ is dependent on the noise of the signal $\varphi(\boldsymbol{x}, t)$. For a typical signal, the selection of different averaging intervals $[t'_1, t'_2]$ results in different levels of smoothing. Cumulative average, defined as follows, can be used to estimate the time span, within which the noise effects are suppressed.

Definition 3.2 The *cumulative time average* of a continuous field variable $\varphi(\boldsymbol{x}, t)$ at a point $\boldsymbol{x} \in \mathbb{R}^3$ and at a time t_0 is defined as:

$$\check{\varphi}(\boldsymbol{x}, t; t_0) := \frac{1}{t - t_0} \int_{t_0}^{t} \varphi(\boldsymbol{x}, t') \, dt' \qquad (3.3)$$

□

The time average of a field variable can alternatively be estimated by function fitting. The selection of the function, which best describes the physical phenomena corresponding to a field variable requires a physical and mathematical approach, which is beyond the scope of this work. However, the weak time dependency of a field variable allows a good estimation of its average with polynomials by using the least squares method (see App. A).

Definition 3.3 A function $\tilde{\varphi}(\boldsymbol{x}, t)$ fitted to a field variable $\varphi(\boldsymbol{x}, t)$ and defined over a time interval $\tau = [t_1, t_2]$ at a point $\boldsymbol{x} \in \mathbb{R}^3$ is called the *estimated time average*. The difference between the field variable and its estimated time average is called the *estimated noise*, $\tilde{\varphi}'(\boldsymbol{x}, t)$, of the field variable:

$$\varphi(\boldsymbol{x}, t) = \tilde{\varphi}(\boldsymbol{x}, t) + \tilde{\varphi}'(\boldsymbol{x}, t) \qquad (3.4)$$

□

If polynomials are used to estimate the average, Theorem A.1 guaranties the suppression of the integral of the estimated noise over the time interval τ:

$$\int_\tau \tilde{\varphi}'(\boldsymbol{x}, t; t_0) dt = 0 \qquad (3.5)$$

Each measurement system has limits within which its amplitude, frequency, and phase responses are constant (Beckwith and Marangoni, 1990). Within the limits of constant responses, a field variable and its corresponding measured values are called consistent:

Definition 3.4 A field variable $\varphi(\boldsymbol{x}, t)$ and its corresponding measured value $\psi(\boldsymbol{x}, t)$ are at a point $\boldsymbol{x} \in \mathbb{R}^3$ and during a time interval $\tau = [t_1, t_2]$ *consistent*, if and only if in their range of variation, their amplitude, frequency, and phase responses are constant. □

The consistency between a field variable and its measured value is dependent not only on the type of the field variable, but also on the measurement system and is generally achieved at low frequencies. The frequency spectrum of noise is generally unknown and most of the devices, which are used for steady state measurements, cannot maintain consistency at noise frequencies. In steady and quasi-steady measurements, however, the time rate of change of the variations is supposed to be small and, therefore, consistency is usually achieved for the mean values.

Definition 3.5 A field variable $\varphi(\boldsymbol{x}, t)$ and its corresponding measured value $\psi(\boldsymbol{x}, t)$ are at a point $\boldsymbol{x} \in \mathbb{R}^3$ and during a time interval $\tau = [t_1, t_2]$ *apparently consistent*, if and only if their time average functions (special or estimated) are consistent. □

With the aforementioned definitions, a quasi-steady variation, as a slight deviation from a steady state, can be defined as follows:

Definition 3.6 A field variable $\varphi(\boldsymbol{x}, t)$ and its corresponding measured quantity $\psi(\boldsymbol{x}, t)$ are at a point $\boldsymbol{x} \in \mathbb{R}^3$ and during a time interval $\tau = [t_1, t_2]$ under *quasi-steady* conditions, if and only if:

1. $\varphi(\boldsymbol{x}, t)$ and $\psi(\boldsymbol{x}, t)$ are apparently consistent.
2. The time rate of change of the time average function (special or estimated) of $\varphi(\boldsymbol{x}, t)$ is bounded.

□

3.3 Zero-drift of pressure sensors

The zero value of pressure sensors in their unloaded state is dependent on the environmental and device conditions, among which dependency on temperature is of major importance (Tropea et al., 2007). After a sensor is set to zero, zero-drift causes gradual change in its output signal, even if the measured quantity is constant.

Figure 3.1 shows typical zero-drifts of five pressure channels measured concurrently during three days. The sensors were set to zero prior to the measurements and were under the same conditions. Besides noise, there is a time variation of the mean, which is not the same among the channels. The variations are within ± 0.003 V corresponding to ± 120 Pa, which is $\pm 0.06\%$ of the measurement range of the sensors in a time interval of 66 hours. This implies the inclusion of zero-drift as an overall change in the measured signal. In this work, zero-drift is considered as a non-negative constant δ_0, and is used to expand both limits of the minimum-maximum range of the variation of a signal. As shown in the figure, the effect of zero-drift depends on the duration of measurement.

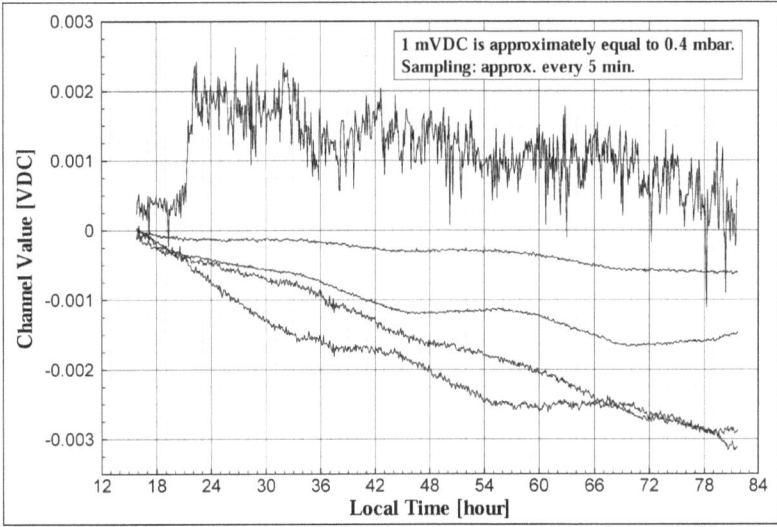

Fig. 3.1: Typical zero-drifts of five pressure channels measured concurrently during 66 hours

3.4 Calibration of local measurements

Measurement devices such as pressure probes and thermocouples measure a physical quantity locally. In the method of calibration by comparison, a *measurement system*, e.g. a pressure measurement loop, is compared with a *reference system*, which is usually a measurement system with higher stability and accuracy. Corresponding to a field variable such as pressure or temperature, the measured quantity by a reference system at a location $x \in \mathbb{R}^3$ and at a time $t \in \mathbb{R}$ is denoted by $\varphi(x, t)$ and that of a measurement system by $\psi(x, t)$. The measured variables φ and ψ can be different physical quantities. For instance, in a pressure field, the measurement system can measure electric potential corresponding to the pressure measured by a reference system.

During the calibration for steady or quasi-steady measurements, typically a subdivision of N_1 points in the common range of the measurement and reference systems, called *data-points*, is selected. In the method used in this work, at each data-point i, $i \in [1, N_1]_\mathbb{N}$, and under the steady or quasi-steady conditions, N_2 values $\psi_i(x, t_j)$, $j \in [1, N_2]_\mathbb{N}$, following a reading of the field variable $\varphi_i(x, t_0)$, $t_0 < t_1$, are registered. The set $\{\psi_i(x, t_j) | j \in [1, N_2]_\mathbb{N}\}$ corresponding to $\varphi_i(x, t_0)$ is called a *sample*.

This method in which one reference value corresponds to multiple readings of a measurement system is called *single-point – multi-point*. In this method the uncertainty parameters of the measurement system are determined by the statistical analysis of the samples but they should be known for the reference system.

If multiple readings are registered for a reference system, so that multiple readings of the reference system correspond to multiple readings of the measurement system at a data-point, the calibration is called *multi-point – multi-point*. In this case, the samples of the reference and the measurement systems correspond to each other at a time t_0 and the goal of data analysis is to provide the measurement and the uncertainty data at this time. If the readings of the reference and measurement systems are temporally concurrent, there will be full or one-to-one correspondence between the sample elements. Therefore, multi-point – multi-point measurements may be *asynchronous* or *synchronous*.

Digital measurement systems assign a discrete output to a continuous input. For example, to a continuous physical quantity $\varphi(\boldsymbol{x}, t) \in D_1 \subset \mathbb{R}$, a discrete value $\psi^d(\boldsymbol{x}, t) \in D_2^d \subset \Psi^d$ may be assigned. The superscript "d" is used to denote a discrete variable or quantity. In some measurement systems, Ψ^d can be defined as:

$$\Psi^d = \{k\delta_1 + \delta_2 \in \mathbb{R} \mid k \in \mathbb{Z}, \ \delta_1 \wedge \delta_2 = const. \in \mathbb{R}\} \tag{3.6}$$

The subset D_2^d is bounded and connected. δ_1, which is the difference between two successive members of Ψ^d, is the *resolution* of $\psi^d(\boldsymbol{x}, t)$.

If a measured quantity ψ_i and its corresponding field variable φ_i are apparently consistent, then the dependency of their average functions or mean values to each other is independent of time. The signal noise is, however, time dependent and is supposed to be stochastic. It can be assumed that the statistical characteristics of the noise such as its probability density function, minimum, maximum, and standard deviation are time independent. Therefore, the correspondence between the mean and the statistical characteristics of the field variable and its corresponding physical quantity will be independent of time. In this way, time independent estimations of the mean value and the statistical quantities are achieved from the time dependent measurements. The goal of the calibration is to make estimations for the mean values as well as the statistical properties of a measured quantity and its field variable at some common time t_0, at which their correspondence is valid.

Figure 3.2 shows a typical sample of 100 values registered from a pressure transducer under quasi-steady conditions. The sample data were measured at equal time intervals. The indices of sample values are shown on the abscissa. The measured quantity is the electric potential. A third order polynomial fit shows an estimated average of the sample. The cumulative time average $\breve{\psi}_i^d(\boldsymbol{x}, t_j; t_0)$, where t_0 corresponds to the first sample value, shows that the noise effect is suppressed approximately after the 10th sample value.

Since the special average, Eq. (3.1), of the sample values of a quasi-steady measurement is not constant in general, the following statistical calculations are based on the estimated time average referred to as

Chapter 3. Theory of Local Measurements under Quasi-Steady Conditions

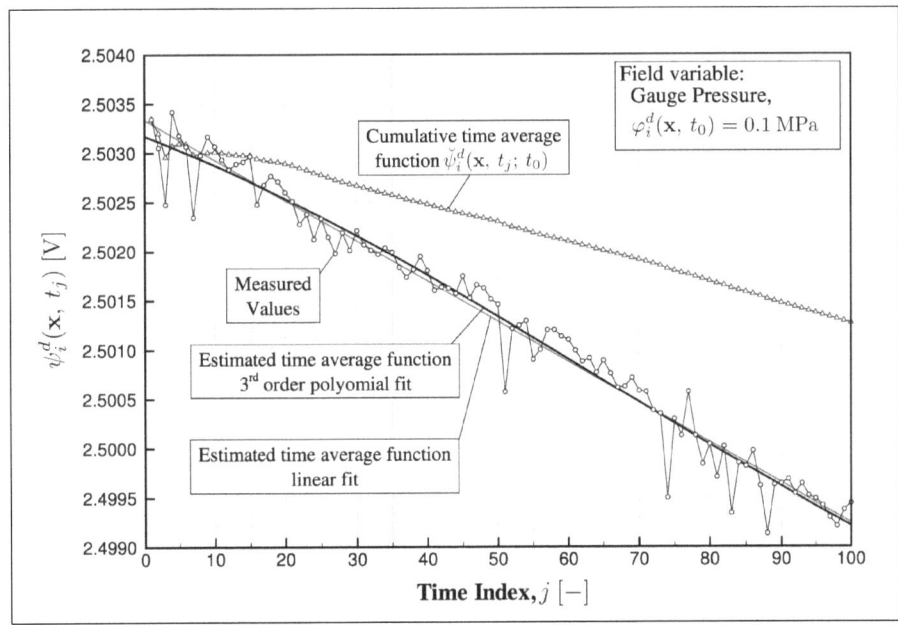

Fig. 3.2: A typical calibration sample of a pressure transmitter
The data were registered at constant time intervals.

the *mean function* and defined as follows:

Definition 3.7 An estimated time average of a bounded sample (a sample with finite elements) is called a *mean function of sample*. For an unbounded sample, it is called the *mean function*. □

The method of least squares is the curve fitting method used for the uncertainty analysis in this work. Details of fitting polynomials with this method are given in App. A.

Definition 3.7 is rather general and does not define the mean function to be the best fit to a sample. In other words, the mean function does not necessarily reveal the physical nature of a measured quantity. As an example, the third order polynomial fit, as depicted in Fig. 3.2, matches the sample better than the straight line fit. The increase in the order of the polynomial results in a better fit to the sample values, but can lead to a departure from the physical phenomena due to the effect of the signal noise.

Since the nature of the physical variation of a quantity and that of the sample noise are not known, the determination of the best fit requires a more sophisticated approach, which accounts for the physical phenomena being investigated. However, small and low rate deviations from a steady state can be approximated well by the low order terms of the Taylor series expansions of the relevant field variables. This implies that low order polynomials could be used to approximate such deviations.

Using the mean function as basis, the statistical quantities are defined as follows:

Definition 3.8 For a sample $\psi_i^d(\boldsymbol{x}, t_j) \in D^d$, $j \in [1, N_2]_{\mathbb{N}}$, at a data-point i with a mean function $\tilde{\psi}_{i, N_2}(\boldsymbol{x}, t)$ over the time interval $\tau \in [t_1, t_2]$, the *probability density function (pdf) of sample* is defined as:

$$f_{i, N_2}^d(\zeta) : \mathbb{R} \longmapsto \mathbb{I}$$

$$f_{i, N_2}^d(\zeta) := \frac{1}{N_2} \mathcal{N}_{j=1}^{N_2} \left(\psi_i^d(\boldsymbol{x}, t_j) \, \Big| \, [\zeta]_{D^d} = \left[\psi_i^d(\boldsymbol{x}, t_j) - \tilde{\psi}_{i, N_2}(\boldsymbol{x}, t_j) \right]_{D^d} \right) \quad (3.7)$$

For an unbounded sample the *probability density function* is:

$$f_i^d(\zeta) := \lim_{N_2 \to +\infty} f_{i, N_2}^d(\zeta) \quad (3.8)$$

□

Definition 3.9 For a sample $\psi_i(\boldsymbol{x}, t_j)$, $j \in [1, N_2]_{\mathbb{N}}$, at a data-point i with a mean function $\tilde{\psi}_{i, N_2}(\boldsymbol{x}, t)$ over the time interval $\tau \in [t_1, t_2]$, the *variance of sample* is defined as:

$$\sigma_{i, N_2}^2 \left(\psi_i(\boldsymbol{x}, t_j) \right) := \frac{1}{N_2} \sum_{j=1}^{N_2} \left(\psi_i(\boldsymbol{x}, t_j) - \tilde{\psi}_{i, N_2}(\boldsymbol{x}, t_j) \right)^2 \quad (3.9)$$

where $\sigma_{i, N_2}(\psi_i(\boldsymbol{x}, t_j)) \in \mathbb{R}^{\oplus}$ is the *standard deviation of sample*. For an unbounded sample the *standard deviation* is defined as:

$$\sigma_i (\psi_i(\boldsymbol{x}, t_j)) := \lim_{N_2 \to +\infty} \sigma_{i, N_2}(\psi_i(\boldsymbol{x}, t_j)) \quad (3.10)$$

□

In order to determine the limits of the variation of a signal, the minimum and maximum deviations of sample from its mean are defined as follows:

Definition 3.10 For a sample $\psi_i(\boldsymbol{x}, t_j)$, $j \in [1, N_2]_{\mathbb{N}}$, at a data-point i with a mean function $\tilde{\psi}_{i, N_2}(\boldsymbol{x}, t)$ over the time interval $\tau \in [t_1, t_2]$, the *minimum and maximum deviation from the mean of sample* are defined as follows, respectively:

$$\check{\epsilon}_{i, N_2}(\psi_i(\boldsymbol{x}, t_j)) := \min_{[1, N_2]_{\mathbb{N}}} \left(\psi_i(\boldsymbol{x}, t_j) - \tilde{\psi}_{i, N_2}(\boldsymbol{x}, t_j) \right) \quad (3.11)$$

$$\hat{\epsilon}_{i, N_2}(\psi_i(\boldsymbol{x}, t_j)) := \max_{[1, N_2]_{\mathbb{N}}} \left(\psi_i(\boldsymbol{x}, t_j) - \tilde{\psi}_{i, N_2}(\boldsymbol{x}, t_j) \right) \quad (3.12)$$

Chapter 3. Theory of Local Measurements under Quasi-Steady Conditions

Table 3.1: Summary of the statistical analysis in the quasi-steady calibration

	Reference System	Measurement System
Mean function of sample	$\tilde{\varphi}(\tilde{\psi})$	$\tilde{\psi}$
Standard deviation of sample	$\sigma_r(\tilde{\varphi})$	$\sigma(\tilde{\psi})$
Minimum deviation from the mean of sample	$\check{\epsilon}_r(\tilde{\varphi})$	$\check{\epsilon}(\tilde{\psi})$
Maximum deviation from the mean of sample	$\hat{\epsilon}_r(\tilde{\varphi})$	$\hat{\epsilon}(\tilde{\psi})$

and for an unbounded sample:

$$\check{\epsilon}_i\left(\psi_i(\boldsymbol{x},\,t_j)\right) := \lim_{N_2 \to +\infty} \check{\epsilon}_{i,\,N_2}\left(\psi_i(\boldsymbol{x},\,t_j)\right) \tag{3.13}$$

$$\hat{\epsilon}_i\left(\psi_i(\boldsymbol{x},\,t_j)\right) := \lim_{N_2 \to +\infty} \hat{\epsilon}_{i,\,N_2}\left(\psi_i(\boldsymbol{x},\,t_j)\right) \tag{3.14}$$

□

The time dependency of a quasi-steady variable is generally unknown. Slight deviations from the steady-state conditions due to slow variations such as a small leakage in a pressure measurement loop or a small heat transfer rate in a temperature measurement loop, are usually nearly linear and polynomials of small orders can be used as the mean functions of sample.

3.5 Correlation functions

The statistical analysis at each data-point $i \in [1,\,N_1]_\mathbb{N}$ results in the estimated mean function, standard deviation, minimum, and maximum of sample for a measured variable.

A reference system is usually calibrated according to the national standards and the results of its statistical analysis, including the uncertainty data, are provided in the calibration certificates.

The combination of the data-points results in the correlation functions of calibration, which provide the dependency between the field variables and their measured values in a measurement system.

Table 3.1 summarizes one way of the combination of the statistical functions of the reference and the measurement systems. A subscript "r" is used to indicate a parameter of the reference system. The reference quantity is a function of the measured quantity, $\tilde{\varphi} = \tilde{\varphi}(\tilde{\psi})$. This function is assumed to be one-to-one and smooth, as is expected from a measurement system. Other statistical quantities are described as functions of their corresponding mean function.

The correlation functions can be generated by curve fitting a set of data-points. Spline interpolation has the property to preserve the mean and the uncertainty values at the data-points. A cubic spline fit provides piecewise-continuous third order polynomials, which are smooth at the data-points. In this work, the third order polynomial is used to approximate the mean function of sample, as shown in Fig. 3.2, and the natural cubic spline is used as the correlation function for the functions in Table 3.1. The mathematical details for function fitting by splines are given in App. B.

Zero-drift is among the effects, which are independent from the calibration data and, therefore, should be superposed on the measurement results. The effect of zero-drift can be minimized during calibration by minimizing the calibration time and stabilizing the device and environmental conditions. As mentioned in Sect. 3.3, zero-drift is considered as a constant $\delta_0 \in \mathbb{R}^\oplus$, which extends the range of the variation of a signal as $\left[\tilde{\psi} + \check{\epsilon}(\tilde{\psi}) - \delta_0,\ \tilde{\psi} + \hat{\epsilon}(\tilde{\psi}) + \delta_0\right]$, in which the location of measurement, x, has been omitted for simplicity.

The internal error of the measurement and the data/signal processing devices is dependent on the device characteristics such as sensitivity,stability, and dependency on the environmental conditions. It is usually determined and is provided by the manufacturer. Device error is included in the calibration data of a measurement loop. However, it determines the limit of the maximum achievable accuracy of the measurements. Considering $\check{\delta}_1(\tilde{\psi})$ and $\hat{\delta}_1(\tilde{\psi})$ as the minimum and maximum values of the device internal errors for $\tilde{\psi}$, respectively, the range of the variation of a field variable ψ will be:

$$\left[\tilde{\psi} + \check{\Delta}(\tilde{\psi}),\ \tilde{\psi} + \hat{\Delta}(\tilde{\psi})\right] := \left[\tilde{\psi} + \check{\delta}_1(\tilde{\psi}),\ \tilde{\psi} + \hat{\delta}_1(\tilde{\psi})\right] \cup \\ \left[\tilde{\psi} + \check{\epsilon}(\tilde{\psi}) - \delta_0,\ \tilde{\psi} + \hat{\epsilon}(\tilde{\psi}) + \delta_0\right] \qquad (3.15)$$

where $\check{\Delta}(\tilde{\psi})$ and $\hat{\Delta}(\tilde{\psi})$ are the minimum and maximum overall deviations of $\tilde{\psi}$, respectively.

3.6 Analysis of measured data

Figure 3.3 presents the dependency between the measurement and the reference systems. The function $\tilde{\varphi}(\tilde{\psi})$ prepares the dependency between the measured quantity $\psi(x)$ and the reference quantity $\varphi(x)$.

For each measured value $\psi(x)$, a set D of the mean values is found, which contains $\psi(x)$ in its range of variation:

$$\mathrm{D} := \left[\tilde{\psi}_1(x),\ \tilde{\psi}_2(x)\right] = \left\{\tilde{\psi}\mid \psi(x) \in \left[\tilde{\psi} + \check{\Delta}(\tilde{\psi}),\ \tilde{\psi} + \hat{\Delta}(\tilde{\psi})\right]\right\} \qquad (3.16)$$

D is the set of all possible mean values, for which $\psi(x)$ might have been encountered during the measurements.

If several measurements $\overset{i}{\psi}(x)$, $i \in [1, M]_\mathbb{N}$, are available for the same physical conditions, then the arithmetic mean $\bar{\psi}(x) = \sum_{i=1}^{M} \overset{i}{\psi}(x)/M$ is considered to be the measured value and its corresponding

Chapter 3. Theory of Local Measurements under Quasi-Steady Conditions 71

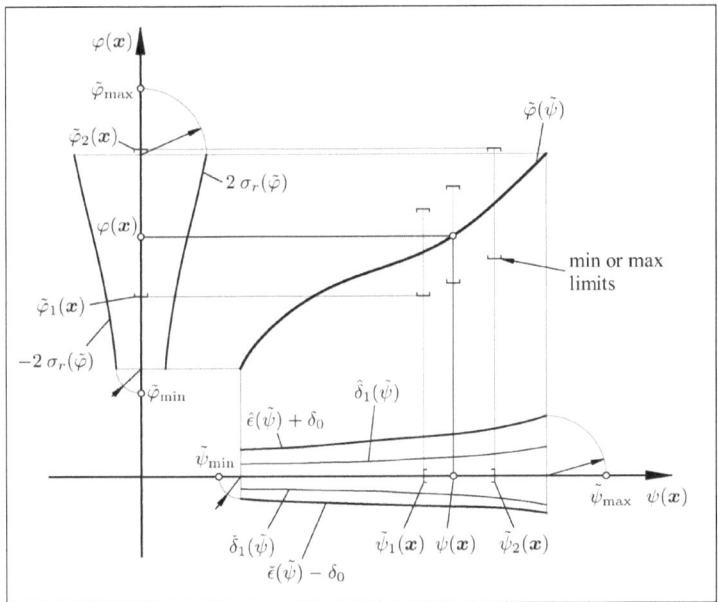

Fig. 3.3: The calibration correlation function and the dependency of the uncertainty parameters

range of variation is determined from each individual range $\overset{i}{D} = [\overset{i}{\tilde{\psi}_1}(\boldsymbol{x}), \overset{i}{\tilde{\psi}_2}(\boldsymbol{x})]$ as follows:

$$D = \left[\bar{\tilde{\psi}}_1(\boldsymbol{x}), \bar{\tilde{\psi}}_2(\boldsymbol{x})\right] = \left[\frac{1}{M}\sum_{i=1}^{M} \overset{i}{\tilde{\psi}}_1(\boldsymbol{x}), \frac{1}{M}\sum_{i=1}^{M} \overset{i}{\tilde{\psi}}_2(\boldsymbol{x})\right] \quad (3.17)$$

For the pressure measurements in this work, the range of variation for the reference quantity is given in the form of (Kalibrierschein DPI 610, 2005):

$$\left[\tilde{\varphi} + \check{\Delta}(\tilde{\varphi}), \tilde{\varphi} + \hat{\Delta}(\tilde{\varphi})\right] = [\tilde{\varphi} - 2\,\sigma(\tilde{\varphi}), \tilde{\varphi} + 2\,\sigma(\tilde{\varphi})] \quad (3.18)$$

Corresponding to the variation range of the measured quantity, D, a range of variation for the reference quantity, R, is found as follows:

$$\begin{aligned} R &:= [\tilde{\varphi}_1(\boldsymbol{x}), \tilde{\varphi}_2(\boldsymbol{x})] \\ &= \{\tilde{\varphi}|\ \psi(\boldsymbol{x}) \in D \wedge \varphi(\psi(\boldsymbol{x})) \in [\tilde{\varphi} - 2\,\sigma(\tilde{\varphi}), \tilde{\varphi} + 2\,\sigma(\tilde{\varphi})]\} \end{aligned} \quad (3.19)$$

R is the variation range of a physical quantity $\varphi(x)$ corresponding to a measured quantity $\psi(x)$ or a set of measured quantities $\overset{i}{\psi}(x)$, $i \in [1, M]_\mathbb{N}$.

3.7 Propagation of error

The combination of the primary or the directly measured quantities in order to calculate the secondary or the derived quantities is accompanied by the combination of the measurement errors. Depending on the sort of combination, different methods can be used to determine the error ranges for the derived quantities. Two methods of error analysis including the analysis of error propagation by the perturbation method, used for the combination of the primary quantities via real valued analytic functions, and the estimation of the error range by numerical computation are presented in this section.

Generally a measured quantity x_i with a deviation $\delta_i \in [\delta_{\min}, \delta_{\max}]$ may experience all of the vales $[x_i + \delta_{\min}, x_i + \delta_{\max}]$. For an ideal measurement, it is expected that $|\delta_i| \ll |x_i|$. In this case, the perturbation parameters $\epsilon_i := \delta_i / x_i \in (-1, 1)$ can be used in asymptotic expansions to determine the resultant of error combinations. In the following examples, this method is applied to the equations of the data analysis of the 5-hole pressure probe measurements of the next chapter.

As the first example, the generalized form of the pressure coefficients in Sect. 4.5 is considered as follows:

$$y = \frac{x_1 + x_2 + x_3}{x_4 + x_5 + x_6} \tag{3.20}$$

Including the deviations and factorizing results in:

$$y + \delta = \left(\frac{x_1 + x_2 + x_3}{x_4 + x_5 + x_6}\right) \left(\frac{1 + \epsilon_1 + \epsilon_2 + \epsilon_3}{1 + \epsilon_4 + \epsilon_5 + \epsilon_6}\right) \tag{3.21}$$

where δ is the deviation of y as the result of the combination of ϵ_i and $\epsilon_i := \delta_i/(x_1 + x_2 + x_3)$ for $i \in [1, 3]_\mathbb{N}$, and $\epsilon_i := \delta_i/(x_4 + x_5 + x_6)$ for $i \in [4, 6]_\mathbb{N}$. If $|\epsilon_i| < 1$, $i \in [1, 6]_\mathbb{N}$, then the second term on the right-hand side can be expanded as follows:

$$\frac{1 + \epsilon_1 + \epsilon_2 + \epsilon_3}{1 + \epsilon_4 + \epsilon_5 + \epsilon_6} = 1 + \epsilon \tag{3.22}$$

where

$$\begin{aligned}
\epsilon := & \epsilon_1 + \epsilon_2 + \epsilon_3 \\
& - \epsilon_4 (\epsilon_1 + \epsilon_2 + \epsilon_3 - \epsilon_4) \\
& - \epsilon_5 (\epsilon_1 + \epsilon_2 + \epsilon_3 - 2\epsilon_4 - \epsilon_5) \\
& - \epsilon_6 (\epsilon_1 + \epsilon_2 + \epsilon_3 - 2\epsilon_4 - 2\epsilon_5 - \epsilon_6) + O^3(\epsilon_1, \ldots, \epsilon_6)
\end{aligned} \tag{3.23}$$

where $O^3(\epsilon_1, \ldots, \epsilon_6)$ stands for all terms of degree three and higher with respect to ϵ_1 to ϵ_6. If the deviations ϵ_i are independent from each other and from x_i, then the variation range of y is determined by the extrema of ϵ, $\epsilon = \delta/y \in [\epsilon_{\min}, \epsilon_{\max}]$. The determination of these extrema becomes complicated as the order of the retained terms of the expansion increases.

The second example is the calculation of velocity from the Mach number, M, and the static temperature of flow, T_{st}, i.e. $V = M \sqrt{\kappa R T_{st}}$, in which κ is the isentropic exponent (and is assumed constant) and R is the gas constant. This equation can be written in a general form as:

$$y = k\, x_1 \sqrt{x_2} \tag{3.24}$$

where k is a real constant. Including the deviations as above, Eq. (3.24) becomes:

$$y + \delta = k\, x_1 \sqrt{x_2} \cdot (1 + \epsilon_1) \sqrt{1 + \epsilon_2} \tag{3.25}$$

From which:

$$\epsilon := \frac{\delta}{y} = \epsilon_1 + \frac{1}{2}\epsilon_2 + \frac{1}{2}\epsilon_1 \epsilon_2 - \frac{1}{8}\epsilon_2^2 + O^3(\epsilon_1, \epsilon_2) \tag{3.26}$$

The next example is the calculation of the static temperature from:

$$T_{st} = T_0 \left(\frac{p_{st}}{p_0} \right)^{\frac{\kappa-1}{\kappa}} \tag{3.27}$$

where the subscripts "st" and "0" denote the static and the stagnation conditions, respectively. Considering the general form:

$$y = x_1 \left(\frac{x_2}{x_3} \right)^k, \quad (k \text{ is a real constant.}) \tag{3.28}$$

and including the deviations, Eq. (3.28) becomes:

$$y + \delta = x_1 \left(\frac{x_2}{x_3} \right)^k \cdot (1 + \epsilon_1) \left(\frac{1 + \epsilon_2}{1 + \epsilon_3} \right)^k \tag{3.29}$$

From which:

$$\begin{aligned}\epsilon := \frac{\delta}{y} &= \epsilon_1 + k(\epsilon_2 - \epsilon_3) + k\,\epsilon_1\,\epsilon_2 + \frac{1}{2}k(k-1)\epsilon_2^2 \\ &\quad - k\,\epsilon_1\,\epsilon_3 - k^2\,\epsilon_2\,\epsilon_3 + \frac{1}{2}k(k+1)\epsilon_3^2 + O^3(\epsilon_1, \epsilon_2, \epsilon_3)\end{aligned} \tag{3.30}$$

The analytical analysis of the error combinations in the above examples for the first order terms is straight forward. With the increase of the order of the retained terms in the expansions, the analytical solutions become difficult. This is also the case, if the combination of the measured quantities is complicated, implicit, or is given by multiple functions. An example is the analysis of the error propagation of the pressure probe measurements in Sect. 4.5.

One simple and rather general solution to such problems, is a numerical search in all possible values. Although it requires considerable computational time, its simplicity and generality makes it a method of choice in many applications. In this method, the variation range of each measured quantity $x_i + \delta_i \in [x_i + \delta_{i,\min}, x_i + \delta_{i,\max}]$ is subdivided and a search algorithm determines the a range of the variation of the derived quantity $y + \delta \in [y + \delta_{\min}, y + \delta_{\max}]$ by evaluating all possible combinations of the primary quantities. By refining the subdivisions and repeating the procedure a convergence criterion can be evaluated.

3.8 Conclusions

This chapter provides the theoretical basis for the data analysis of the velocity measurements by the pressure probes in the next chapter. A theory is presented for the calibration under quasi-steady conditions. By including slow signal variations due to external and device instabilities, it provides a method for on-site calibration with higher accuracy than the regular methods.

Based on raw calibration data, a complete data analysis for the steady state local measurements is presented. The method derives the statistical characteristics of the measured quantities by direct manipulation of the calibration data and without the use of the statistical distribution functions.

The theory developed in this chapter is applied to the pressure probe measurements in the next chapter. The numerical search algorithm of the previous section is used to compute the error propagation. The results of the application of the uncertainty analysis method are shown in Fig. 4.9 (primary quantities) and Fig. 4.10 (derived quantities). The calibration charts of the pressure measurement loops are presented in App. G.

Chapter 4

Velocity Measurement at the Compressor Inlet by Pressure Probes

4.1 The test facility

The centrifugal compressor test facility at the Institute of Turbomachinery and Fluid Dynamics (TFD) is composed of a 1.35 MW motor-driven single stage centrifugal air compressor capable of operating in closed- and open-loop modes. Figure 4.1 is a simplified model of the compressor and Fig. 4.2 presents the process flow diagram of the test facility. A summary of the characteristics and performance data of the test facility is given in Tab. 4.1.

The compressor inlet consists of inlet guide vanes (IGVs) with a hub and a connecting duct between the IGV and impeller casing, which are designed and optimized for generating positive preswirl in the inlet flow. The IGV consists of s-cambered profiles, which are adjustable at $-20°$, $0°$, $20°$, $40°$, $60°$, and $75°$ independently. The vane carrier ring provides continuous circumferential adjustments of the IGV. More information about the inlet section can be found in Hagelstein et al. (2001), Seume et al. (2007), Van den Braembussche et al. (2006), and Mohseni et al. (2010).

4.2 Pressure and temperature measurement

Figure 4.3 shows the block diagram of the pressure and temperature measurements. The pressure and temperature signals are acquired by an HP 34970A data acquisition/switch device (Hewlett-Packard, 1997) and are sent to a computer.

Temperature is measured by nickel-chromium/nickel-aluminum, also known as chromel/ alumel, thermocouples categorized as type K in IEC[1] 584-1, 1995. The type K thermocouple is the most commonly used thermocouple with the specified operating range of -200 °C to 1100 °C (up to 1300

---

[1] IEC: International Electrotechnical Commission

Chapter 4. Velocity Measurement at the Compressor Inlet by Pressure Probes

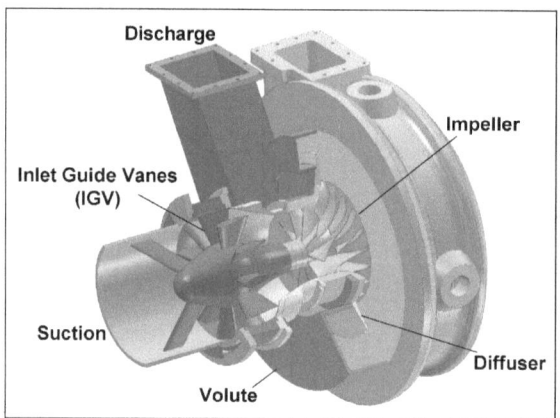

Fig. 4.1: The simplified schematic diagram of the compressor with the IGV

Table 4.1: The characteristics and performance data of the compressor test facility

Medium:	Air
Operating mode:	open-loop, closed-loop
Operating speed:	5000 – 18000 rpm (520 – 1880 rad/s)
Max. design speed:	29000 rpm (3030 rad/s)
Impeller inlet hub diameter:	90 mm
Impeller inlet tip diameter:	280 mm
Impeller outlet diameter:	400 mm
Drive:	DC-Motor, 1350 kW, 40 – 1800 rpm (4 – 188 rad/s)
Max. total pressure ratio:	2.35
Max. isentropic efficiency based on total values:	83 %
Max. corrected mass flow rate:	9.5 kg/s
Inlet Reynolds number:	$2.4 \times 10^5 - 1.6 \times 10^6$
Corrected mass flow rate at the best operating point:	5.5 kg/s
Total pressure ratio at the best operating point:	1.75
Corrected shaft speed at the best operating point:	14500 rpm (1520 rad/s)

Chapter 4. Velocity Measurement at the Compressor Inlet by Pressure Probes   77

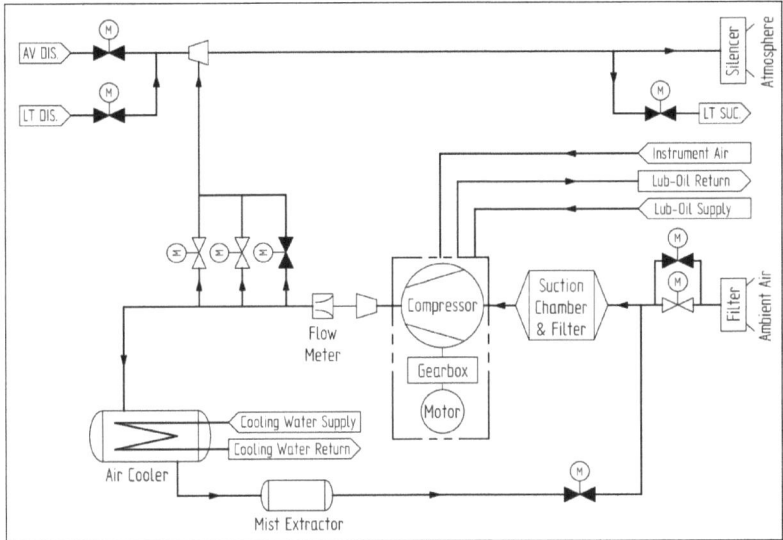

Fig. 4.2: The process flow diagram of the compressor test facility, with valves positioned for the open-loop operation

°C for short duration readings). Its emf[1]-temperature dependence is nearly linear. The chemical composition of type K is 90% Ni-10% Cr / 95% Ni, balanced with Al, Si, Mn. In Germany, a slightly different composition is used: 85% Ni-12% Cr / 95% Ni-3% Mn, 2% Al, 1% Si (Michalski et al., 2001).

4.3 Zero-drift of pressure sensors

Prior to the measurements, all active pressure channels are set to zero. During the measurements, zero-drift causes gradual deviation of the sensors from their unloaded conditions. Figures 4.4 and 4.5 show the zero-drift of the pressure channels during one day. Deviation from zero is composed of a high frequency variation due to the device and ambient noises and a low rate or gradual drift. Table G.2 in App. G provides a summary of the zero-drift ranges of the pressure channels during 24 hours. In this work, zero-drift is considered as an stochastic phenomenon with symmetric range of variation.

---

[1] emf: electromotive force

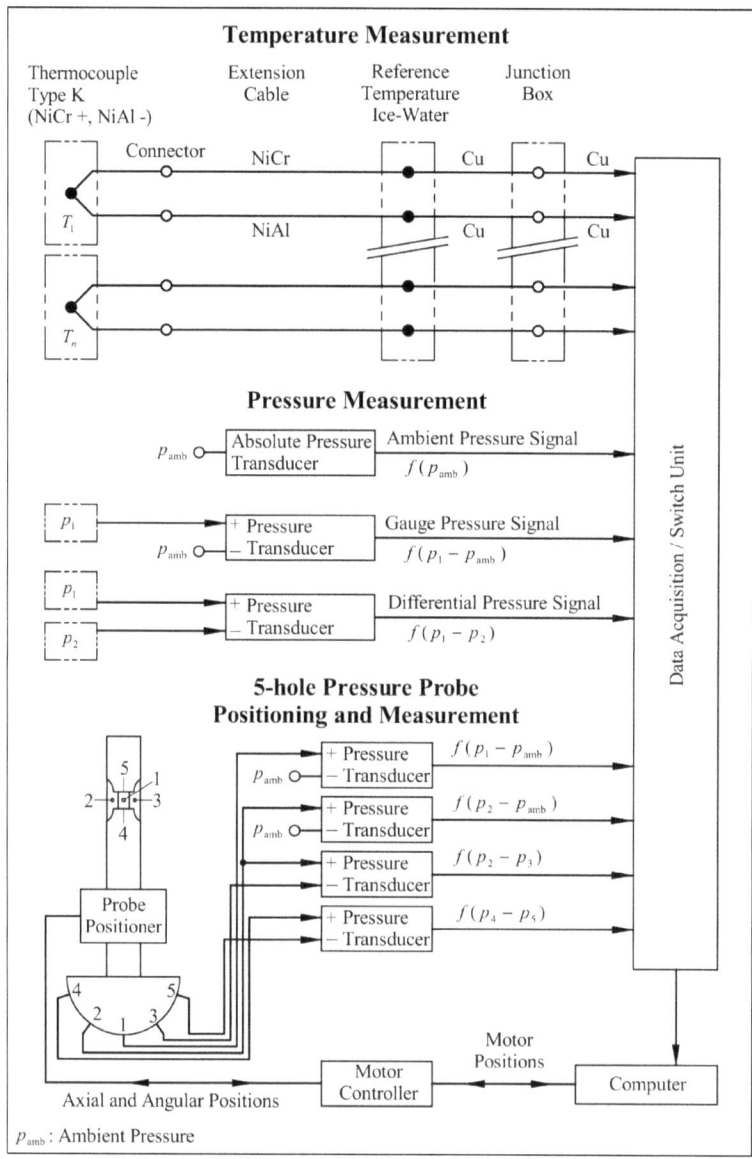

Fig. 4.3: The measurement system diagram

Chapter 4. Velocity Measurement at the Compressor Inlet by Pressure Probes

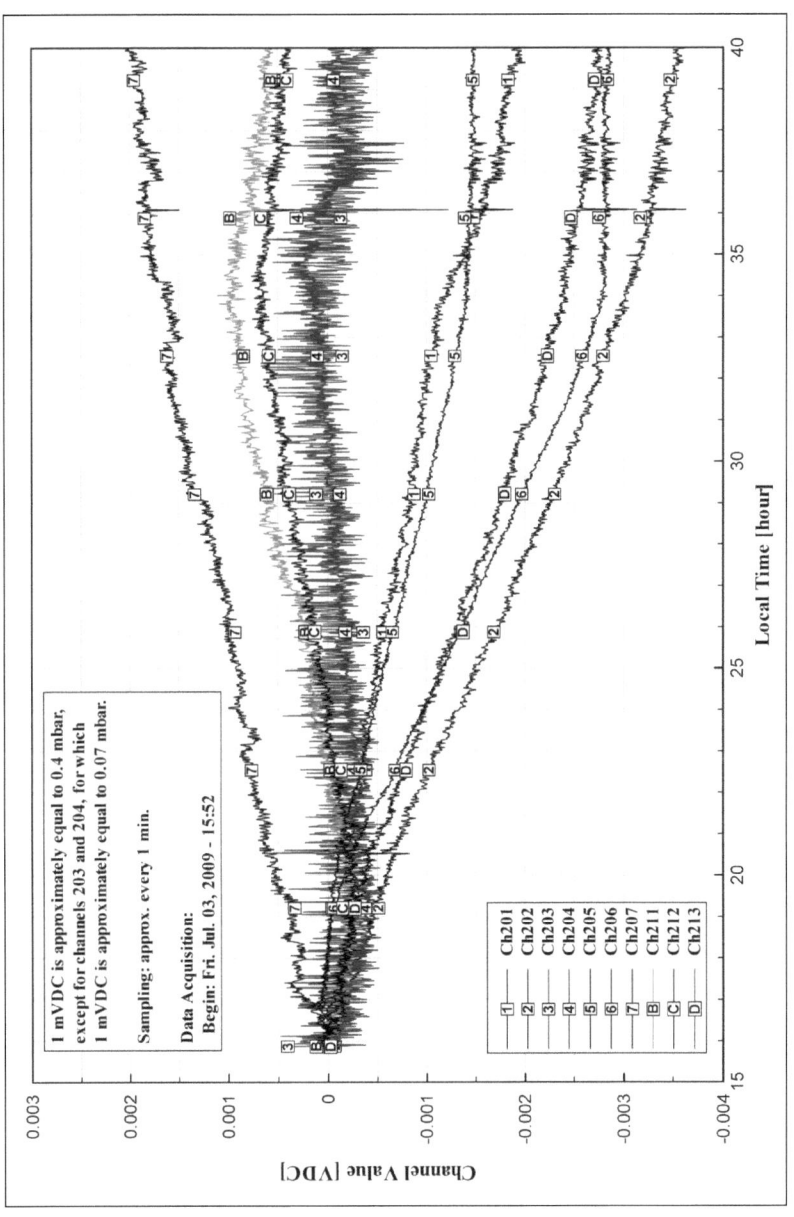

Fig. 4.4: The zero-drift of the pressure transducers during one day (channels 201 to 213)

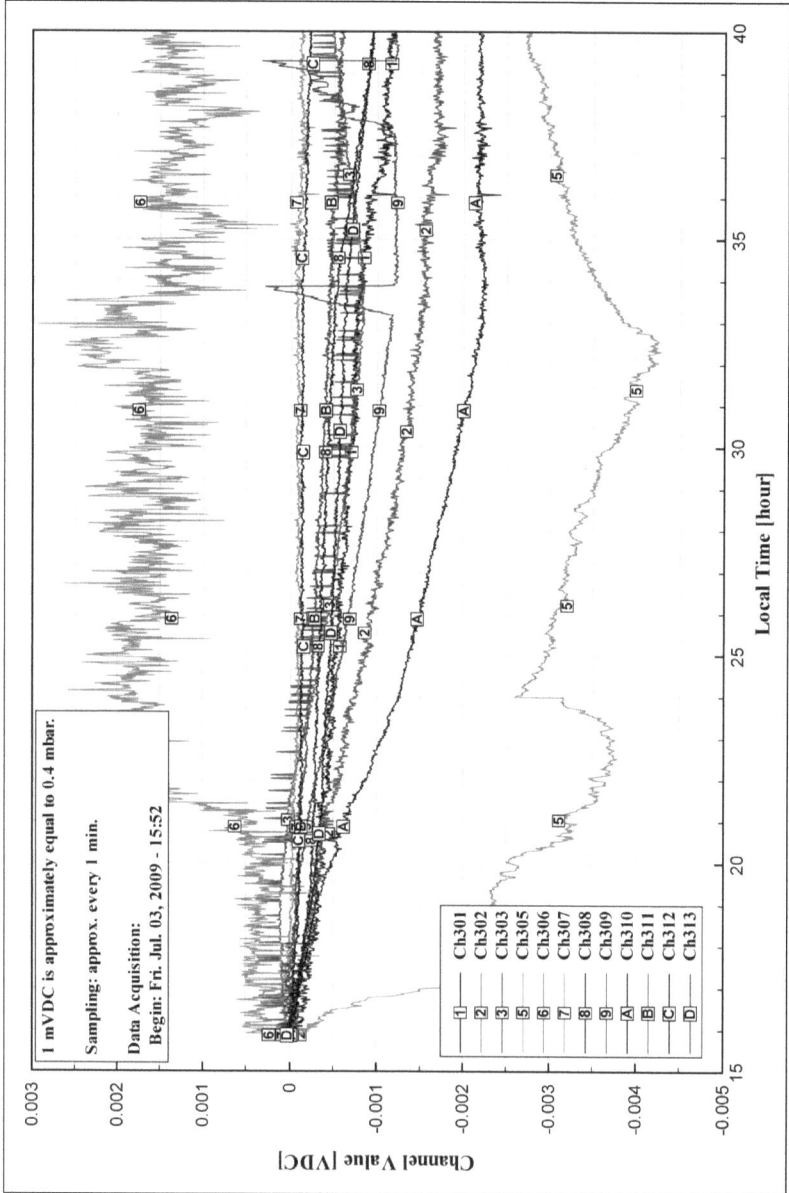

Fig. 4.5: The zero-drift of the pressure transducers during one day (channels 301 to 313)

Chapter 4. Velocity Measurement at the Compressor Inlet by Pressure Probes   81

4.4 Calibration of pressure and temperature measurement loops under quasi-steady conditions

Each pressure and temperature sensor is calibrated in its own measurement loop as shown in Fig. 4.3 by the method of *calibration by comparison*. Differential pressure transducers are calibrated by pressurizing their pressure side, while their vacuum side is open to the ambient. Figure 4.6 characterizes the calibration setups, which are the same as their final setup for the measurements in Fig. 4.3. For each sensor a number of data-points in its operating range, which covers the expected measurement range for the experiments, is selected. In a single-point – multi-point configuration and under quasi-steady conditions, corresponding to a reading from a reference device, a number of readings from the measurement system is registered. Using the statistical analysis of Ch. 3, the mean value and the error estimation of each data-point is determined. These data are combined into calibration curves by function fitting. The calibration correlations provide the basis for the analysis of the measured data. The calibration charts of the pressure transmitters are given in App. G.

4.5 Velocity measurement by 5-hole pressure probes

The flow velocity at the compressor inlet was measured in three sections, Fig. 4.7, with the 5-hole pressure probes, which are capable of determining all three components of the velocity. Each probe enters at the shroud side and travels perpendicular to the machine axis toward the hub in a plane, which passes through the machine axis.

The coordinate system of the pressure probe measurements is defined in Fig. 4.8. For each probe, the reference coordinates is the coordinates, for which the origin is located at the hub along the probe path, the x_1-axis is parallel to the machine axis in the downstream direction (toward the impeller), the x_2-axis is in the radial direction toward the machine axis, and the x_3-axis is in the circumferential direction. The velocity components of a velocity vector $\vec{V}$ in the reference system $x_1\,x_2\,x_3$ are V_1, V_2, and V_3.

Multi-hole pressure probes are used in *calibrated* and *null-reading* modes (Tropea et al., 2007). In calibrated mode, the probe is fixed and the flow direction is determined from the calibration correlations. In this mode, the calibration procedure is detailed and includes rather sophisticated data processing. The advantage is the reduction of the data acquisition time approximately to the time constants of the measurement system, which is considerably faster than that of the null-reading mode.

In the null-reading mode, which is used in this work, each probe is rotated around the x_2-axis, Fig. 4.8b, until p_2 and p_3 become equal. In this way, the measurements are approximately independent from the yaw angle and the calibration time and data reduce considerably. The dependency of the measured data on the yaw angle reduces to a correction for the difference between the aerodynamic yaw angle

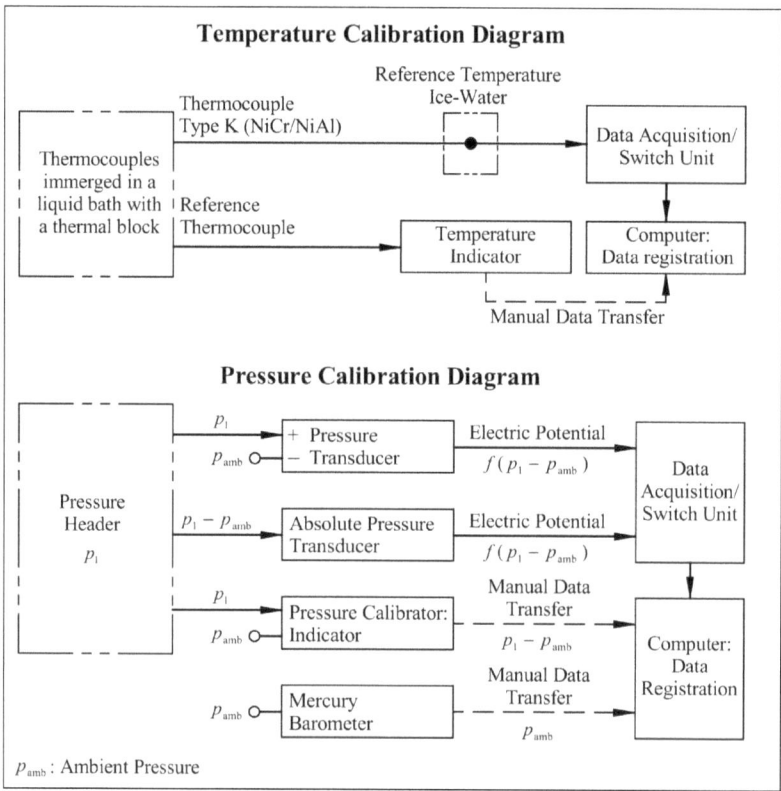

Fig. 4.6: The calibration setup diagrams

(at which $p_2 = p_3$) and the geometric yaw angle (based on the geometric position of the probe), which is correlated as the offset of the yaw angle.

Finding the location, at which the aerodynamic yaw angle vanishes, is dependent on the relaxation times of the measurement system. Therefore, the measurements in null-reading mode are considerably slower than that in the calibrated mode.

In a probe measurement plane, Fig. 4.7, if the probe coordinate system $x_1 x_2 x_3$, Fig. 4.8b, is indicated as $x'_1 x'_2 x'_3$ at its reference position, then its location, at which $p_2 = p_3$, is fully described by a rotation angle α', measured from the positive x'_1-axis in the $x'_1 x'_3$-plane and positive toward the positive x'_3-axis, and a length along the x'_2-axis, which determines the axial position of the probe. At the balance of p_2 and p_3, the yaw angle in the probe coordinates $x_1 x_2 x_3$ is given by the offset of the yaw angle as $-\alpha_0$.

Chapter 4. Velocity Measurement at the Compressor Inlet by Pressure Probes

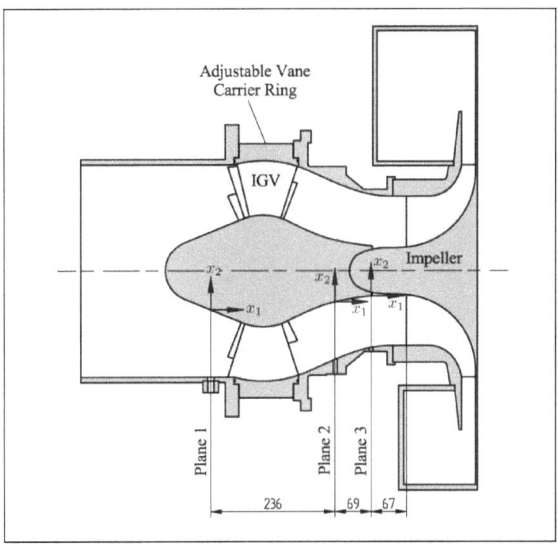

Fig. 4.7: The schematic horizontal section of the compressor through the impeller center line
The three sections (planes 1, 2, and 3) show the locations of the pressure probes. The reference coordinates of the probes, x_1 x_2 x_3, are shown at their origins. For the measurements of this work, the IGV is at zero setting angle and circumferentially located so that the probes measure in the plane midway between two adjacent IGV vanes. The dimensions are in mm.

The yaw angle in the reference coordinates $x'_1 x'_2 x'_3$ will then be $\alpha' - \alpha_0$.

The calibration of each probe in the null-reading mode provides the following correlations:

$$\alpha = \alpha(C_{YA}, M) \tag{4.1}$$
$$\gamma = \gamma(C_{PA}, M) \tag{4.2}$$
$$C_{DP} = C_{DP}(\gamma, M) \tag{4.3}$$
$$C_{SP} = C_{SP}(\gamma, M) \tag{4.4}$$
$$C_{TP} = C_{TP}(\gamma, M) \tag{4.5}$$

where α is the yaw angle, γ is the pitch angle, M is the Mach number, and:

$$C_{YA} := \frac{p_2 - p_3}{p_1 - \bar{p}_{23}} \quad \text{Yaw Angle Coefficient} \tag{4.6}$$

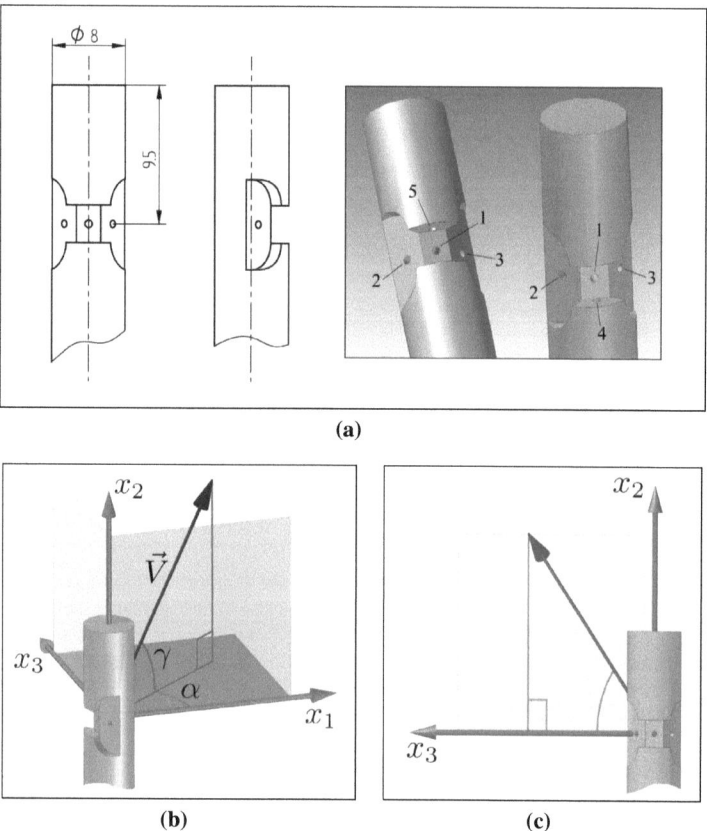

Fig. 4.8: The definition of the pressures and coordinates of the 5-hole pressure probe for measurement in the calibrated mode: **(a)** the 5-hole pressure probe with pressure hole numbering, **(b)** three-dimensional view with positive yaw, α, and pitch, γ, angles of the flow velocity $\vec{V}$, and **(c)** the side view of the coordinates
The x_2-axis coincides with the probe center line. The x_1-axis passes through the pressure hole 1. The coordinate system is attached to the probe and is moved with it. The yaw angle, α, is zero in null-reading mode. The dimensions are in mm.

Chapter 4. Velocity Measurement at the Compressor Inlet by Pressure Probes

$$C_{PA} := \frac{p_4 - p_5}{p_1 - \bar{p}_{23}} \quad \text{Pitch Angle Coefficient} \tag{4.7}$$

$$C_{SP} := \frac{\bar{p}_{23} - p_{st}}{p_1 - \bar{p}_{23}} \quad \text{Static Pressure Coefficient} \tag{4.8}$$

$$C_{TP} := \frac{p_1 - p_0}{p_1 - \bar{p}_{23}} \quad \text{Stagnation Pressure Coefficient} \tag{4.9}$$

$$C_{DP} := \frac{p_1 - \bar{p}_{23}}{p_1} \quad \text{Dynamic Pressure Coefficient} \tag{4.10}$$

$$\bar{p}_{23} := \frac{p_2 + p_3}{2} \tag{4.11}$$

In the above equations, p_{st} is the static pressure, p_0 is the stagnation pressure, and p_n, $n \in [1,5]_{\mathbb{N}}$, is the pressure value corresponding to the n-th pressure hole of the probe, Fig. 4.8a.

According to the calibration data, the yaw angle is approximately independent from the Mach number. The calibration correlations for the probes used in the measurements are given in App. F. The details of the calibration and measurement by pressure probes can be found in Bubolz (2005), Gizzi (2000), and Tropea et al. (2007).

The solution of the correlations (4.2) and (4.3) results in the pitch angle and the Mach number in the probe coordinates. The pitch angle in the probe coordinates $x_1 x_2 x_3$, γ, is measured at $p_2 = p_3$ and the pitch angle in the reference coordinates $x'_1 x'_2 x'_3$, γ' is calculated as follows:

$$\gamma' = \tan^{-1} \left(\frac{\tan \gamma}{\cos(\alpha' - \alpha_0)} \right) \tag{4.12}$$

With the pitch angle and the Mach number, the static and dynamic pressures can be calculated from the correlations (4.4) and (4.5).

The total temperature at the inlet is measured by a half-cylindrical windowed thermocouple. This temperature probe measures the recovery temperature, T_R, of the flow with a recovery factor, r, of approximately 96%–98% (Dillmann et al., 2007):

$$T_R = T_{st} + r \frac{V^2}{2 C_p} = T_{st} + \frac{r(\kappa - 1) V^2}{2 \kappa R} \tag{4.13}$$

From the relation between the static and the stagnation quantities (Wilson and Korakianitis, 1998):

$$T_0 = T_{st} \left(\frac{p_{st}}{p_0} \right)^{\frac{\kappa-1}{\kappa}} \tag{4.14}$$

and considering that $T_0 = T_R$ for $r = 1$, the static temperature and the magnitude of the flow velocity can be calculated as follows:

$$T_{st} = \frac{T_R}{1 - r + r \left(\frac{p_{st}}{p_0}\right)^{\frac{1-\kappa}{\kappa}}} \qquad (4.15)$$

$$V = \sqrt{2\left(\frac{\kappa}{\kappa - 1}\right) R T_R \frac{1 - \left(\frac{p_{st}}{p_0}\right)^{\frac{\kappa-1}{\kappa}}}{r + (1 - r)\left(\frac{p_{st}}{p_0}\right)^{\frac{\kappa-1}{\kappa}}}} \qquad (4.16)$$

These scalar quantities are independent from the coordinates and are the same in both probe and reference coordinate systems.

4.6 Results of velocity measurement by pressure probes

The flow measurement at the inlet of the compressor with 5-hole pressure probes consists of the flow measurement in three planes, Fig. 4.7:

- Upstream of the inlet guide vanes (IGV), plane 1.
- Downstream of the IGV, plane 2.
- Upstream of the impeller, plane 3.

The measurements are performed at the best operating point or the point of maximum efficiency of the compressor, i.e. at the corrected shaft speed of 16000 rpm and at the corrected mass flow rate of 5.87 kg/s. The IGV vanes are at zero setting angle and circumferentially located so that the plane of the pressure probes is midway between two adjacent vanes. Under these conditions, probe pressures in spanwise direction from the hub to the shroud in null-reading mode together with other required quantities, such as the stagnation temperature in the suction chamber, are registered. For each data-point, ten readings are registered in order to suppress the unsteady effects in the flow.

The application of the theory of measurements under the quasi-steady conditions in Ch. 3 to velocity measurement by the 5-hole pressure probes is presented in Figs. 4.9 to 4.11. Figure 4.9 shows the variation of the channel values or the primary quantities of the pressure probes in planes 1 and 2. Although the measurement loops are the same, they show different error levels. The accuracy of the measurements is good enough to capture the flow variations such as the wall effects in the shroud-side. The resulting pressure values and their corresponding error ranges are characterized in Figs. 4.10 and 4.11. The error levels show the difference between the measurement channels and have negligible variation in the spanwise direction. The dependency of the measurement channels and the probe pressures is given in Tab. G.1 in App. G. The method provides the ranges of uncertainty at each data

Chapter 4. Velocity Measurement at the Compressor Inlet by Pressure Probes 87

point, in which the stochastic properties of the channel, the effect of zero-drift, and the result of error propagation are included. It also reveals the relative accuracy of each measurement channel.

Figure 4.12 shows the variation of the static and stagnation pressures in the spanwise direction. The difference between the stagnation pressures in planes 1 and 2 is within the range of their common uncertainty. In the vicinity of the shroud they show the flow losses in the boundary layer, which are developed along the inlet pipe. The resulting velocity magnitudes and components, and the Mach number variation in the spanwise direction in planes 1, 2, and 3 are shown in Figs. 4.13 and 4.14. Since the IGV is at zero setting angle, the difference between the velocity magnitude and its axial component is small. The plausibility of the velocity values were checked with the average velocity values based on the measured mass flow rate.

In the error analysis of the measured data in this chapter, the uncertainty data of the pressure probes is excluded. As a future work, the presented calibration method can be implemented in the calibration of the 5-hole pressure probes, in order to prepare their uncertainty data.

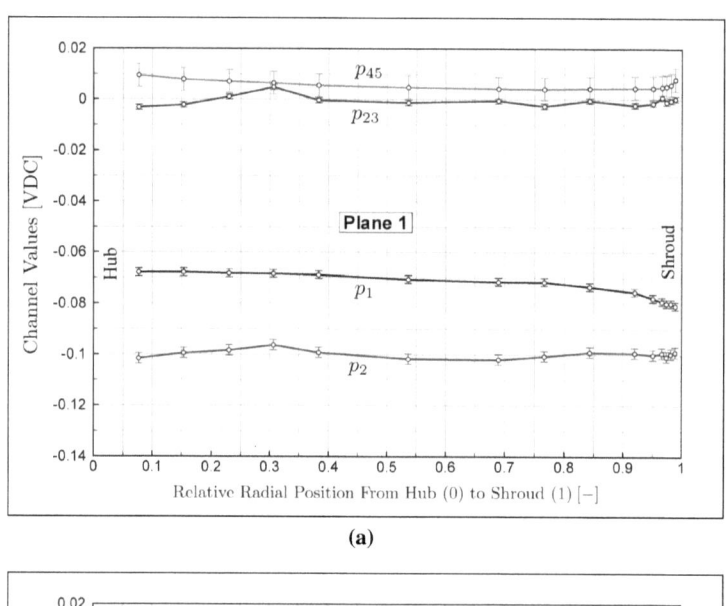

(a)

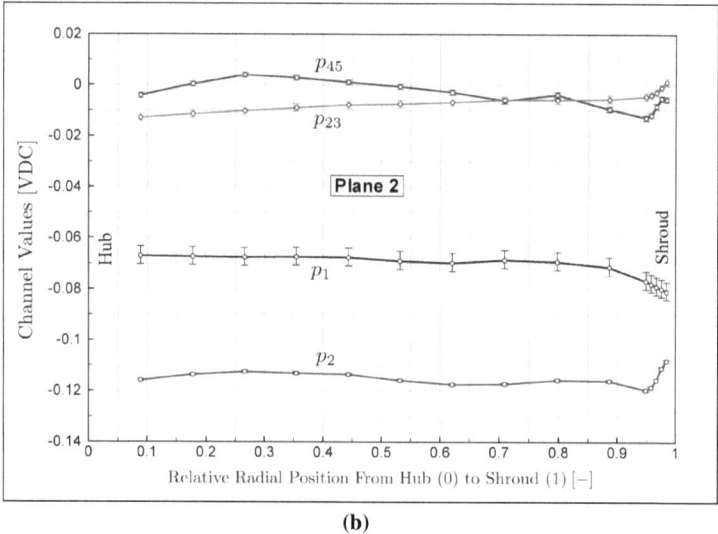

(b)

Fig. 4.9: Typical results of the steady state measurement of the primary quantities

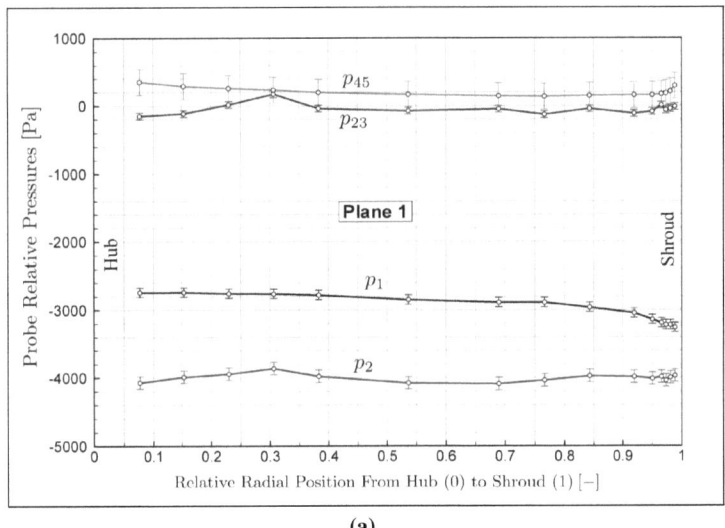

(a)

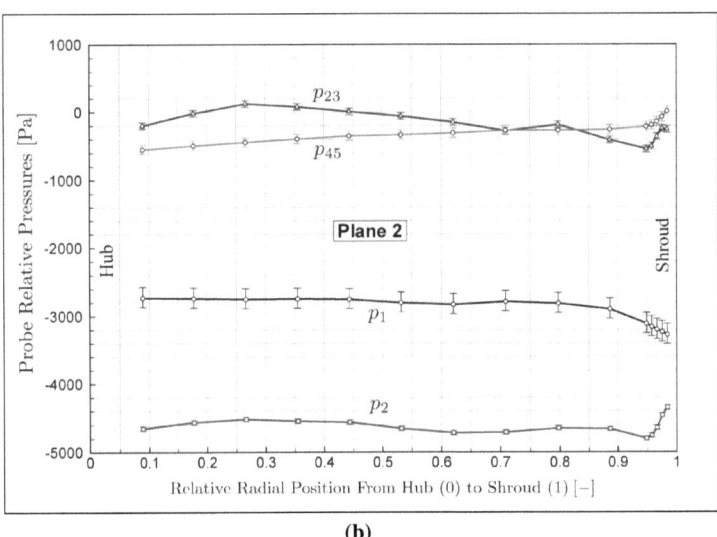

(b)

Fig. 4.10: The corresponding pressure values of Fig. 4.9 after data analysis

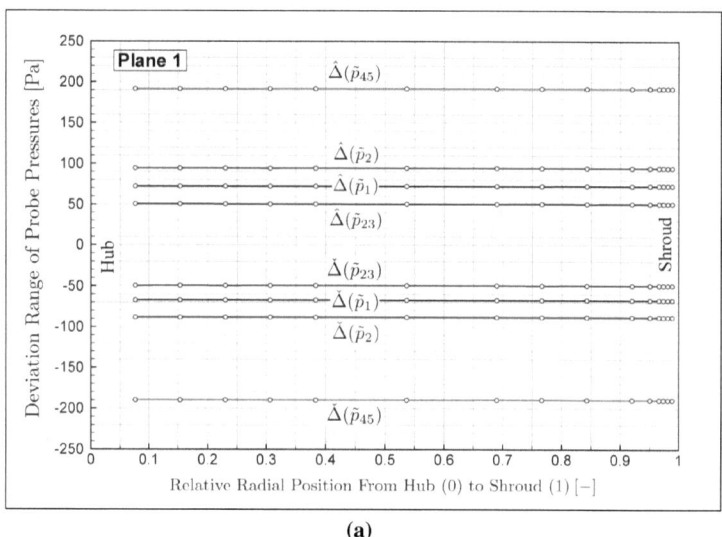

(a)

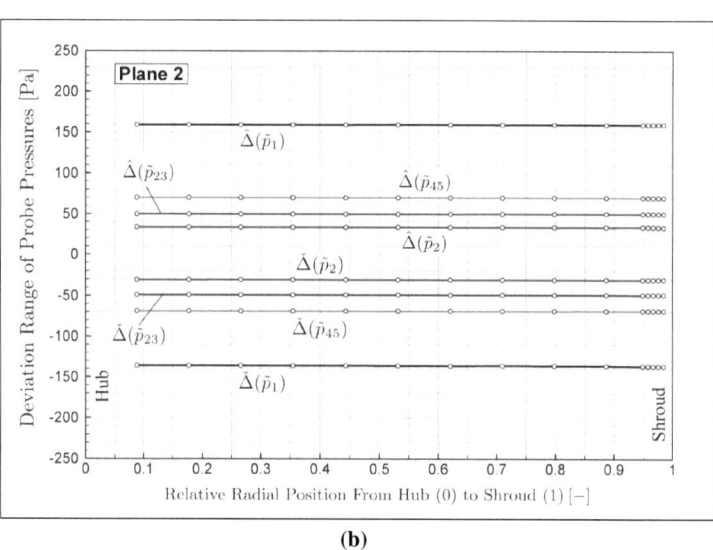

(b)

Fig. 4.11: Error ranges corresponding to the pressure values of Fig. 4.10

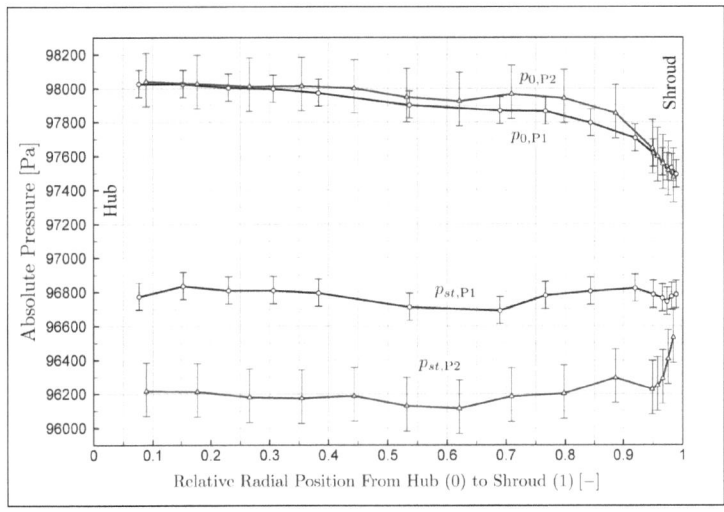

Fig. 4.12: The variation of the static and stagnation pressures in planes 1 and 2 in the spanwise direction
The indices P1 and P2 represent the planes 1 and 2, respectively.

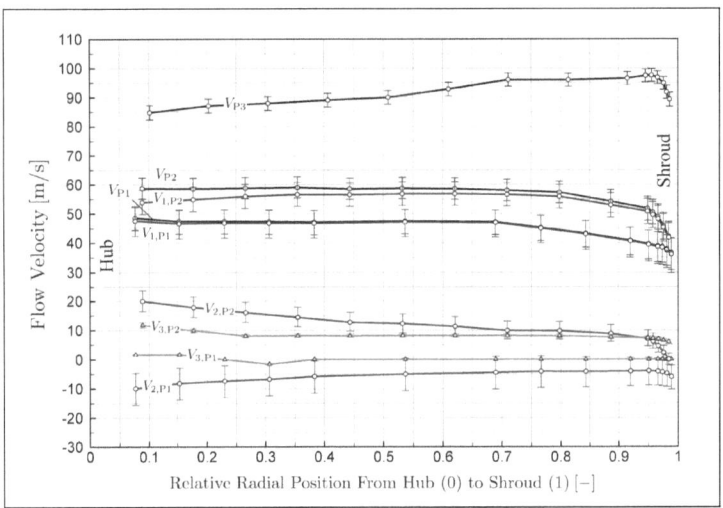

Fig. 4.13: Velocity variations in planes 1, 2, and 3 in the spanwise direction
The indices P1, P2, and P3 stand for the planes 1, 2, and 3, respectively.

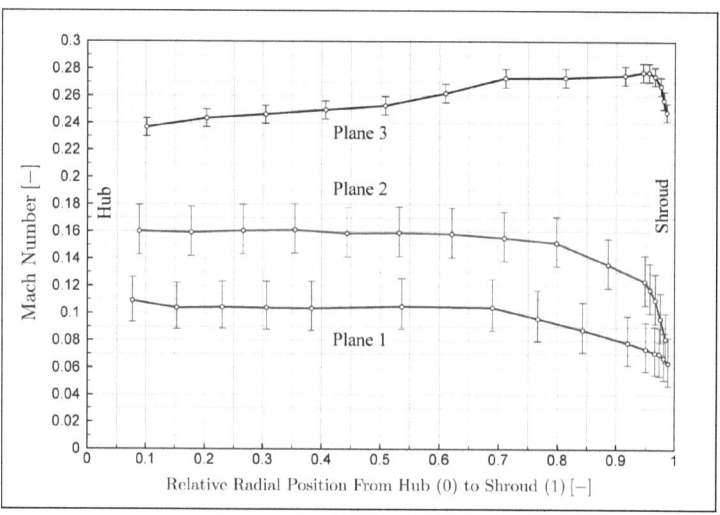

Fig. 4.14: Variation of the Mach number in planes 1, 2, and 3 in the spanwise direction

Chapter 5

PIV Measurement at the Compressor Inlet

In this chapter, the PIV setup and the measurements in the compressor inlet are presented. After an introduction to the setup, the calibration and the back-transformation of the PIV images using the method of Ch. 2 are presented. The practical aspects of the measurements are considered next and are followed by a comparison between the PIV and the pressure probe measurements.

5.1 Endoscopic SPIV setup at the compressor inlet

In particle image velocimetry (PIV) the velocity field in a plane is captured by the analysis of two successive images from the particles, which move with a fluid flow. The plane is illuminated by a laser light-sheet. The particle density and the delay between the images are adjusted so that their local correlation results in a physical velocity field. At each point in the light-sheet, a camera is capable of capturing a projected locally averaged velocity in the plane, which passes through the point and is perpendicular to the line of sight of the camera. For instance, a camera mounted normal to a light-sheet can capture the velocity components in the light-sheet in the vicinity of its optical axis.

In the stereoscopic PIV (SPIV), two cameras, whose lines of sights are not parallel with each other and the light-sheet, are used to capture all components of the velocity vectors. Based on the optics and the geometry of the SPIV setup, the combination of the coincident velocity components captured by each camera, results in the full velocity vector field.

For an oblique angle of view, which is typical in SPIV, in order to have the image in focus the Scheimpflug criterion (Scheimpflug, 1904) should be satisfied. This criterion states that in order to have the image in focus in an optical arrangement, if the object, lens, and image planes are perpendicular to a common plane, then they should intersect at a common line. A comprehensive account of PIV and its applications can be found in Raffel et al. (2007), Tropea et al. (2007) and Westerweel (1993).

Figure 5.1 shows the PIV setup for the flow measurements after the IGV and upstream of the impeller. The laser light from a double oscillator Nd:YAG laser with a nominal power of 200 mJ/Pulse at a wavelength of 532 nm is guided through a laser arm toward the light-sheet endoscope. The light-sheet is generated along the endoscope, i.e. at 0° with respect to the endoscope axis, and its plane passes through the machine axis. The maximum laser power out of the light-sheet endoscope is about 28 mJ/Pulse. With this amount of illumination, the measurement results are very sensitive to the settings and the quality of the setup. DEHS[1] atomized by the Laskin nozzles is used for seeding. The air stream is seeded in the suction chamber.

PIV images are recorded by two Sensicam QE double shutter cameras[2] equipped with cooled 12 Bit CCD chips, capable of recording in spectral range of 290 to 1100 nm. The records are in gray scale with the resolution of 12 Bit, which is 16 times more accurate than that of 8 Bit cameras. The cameras are able to capture 1376×1040 pixel double images with internal interframing time of 500 ns at 10 Hz. External connections and cabling increases the overall interframing time of the setup to about 5 μs.

Two camera endoscopes, with the view angles of 0° with respect to the endoscope axis, at about 45° on the sides of the light-sheet endoscope provide the stereoscopic view of the light-sheet. The cameras and the laser endoscopes are coplanar. Scheimpflug adapters are used to enhance the image quality of the cameras.

A two-dimensional view of the PIV measurement section is shown in Fig. 5.2, in which the geometrical locations of the pressure probes and those of the PIV as well as the coordinates and the location of the PIV data in the flow channel are shown.

The design of the PIV setup provides both direct access to the flow channel, in which the endoscopes are connected to the casing, and indirect access, in which the endoscopes are not connected to the casing and the optical access is provided by small windows. In both cases, the setup provides the ability to clean the optical paths during the measurements. The arrangement of the setup provides the flow investigation near the impeller, while maintaining the background illumination limited. As intended for a research work, the setup favors a simple design with redundant degrees of freedom to facilitate the adjustments.

5.2 Calibration

The calibration grid, its adjusting mechanism and the calibration images of cameras 1 and 2 are shown in Fig. 5.3. The calibration grid is vertical to the plane of the endoscopes and consists of a 2 mm × 2 mm rectangular grid of symmetric Gaussian patterns. Two adjusting screws, allow the adjustment of its position in the horizontal plane with the accuracy of 0.01 mm. At each position the calibration grid can be turned toward each camera. An L marking in the grid locates the common origin or the reference point of the images. Due to manufacturing tolerances, the optical axis of the endoscope may not

---

[1] DEHS: Di-ethyl-hexyl-sebacate
[2] PCO Computer Optics GmbH. Donaupark 11, 93309 Kelheim, Germany. URL: www.pco.de.

Chapter 5. PIV Measurement at the Compressor Inlet 95

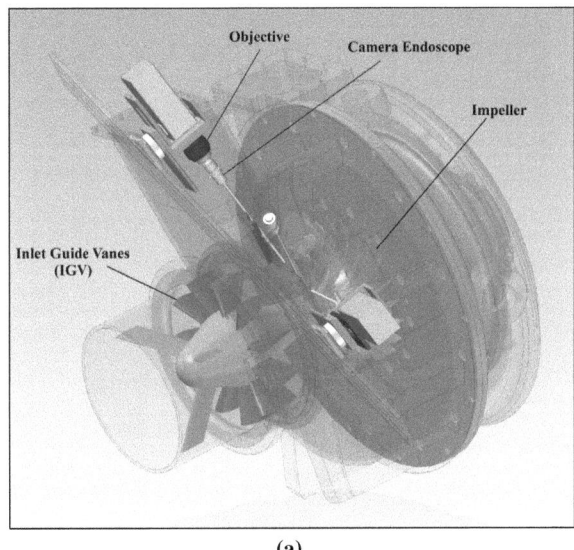

(a)

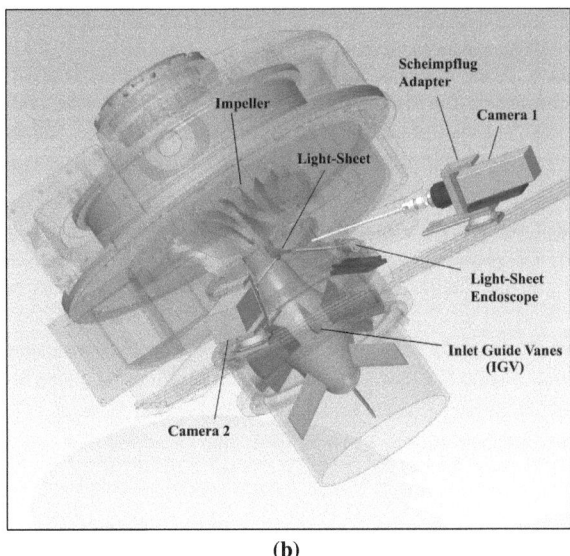

(b)

Fig. 5.1: The endoscopic stereoscopic PIV setup on the compressor for flow measurements between the IGV and the impeller

96   Chapter 5. PIV Measurement at the Compressor Inlet

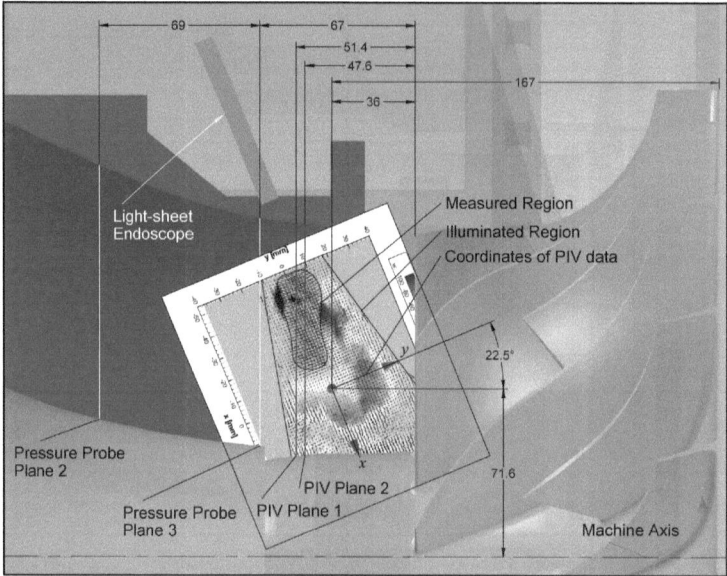

Fig. 5.2: The PIV measurement section
Shown are the illuminated and measured regions and the location of the pressure probe planes. The PIV planes 1 and 2 are the locations of the PIV data extraction in Fig. 5.11 for comparison with the pressure probe data. The measured region is limited to a subregion of the illuminated area. The dimensions are in mm.

match its geometrical axis. This is seen as a downward shift of the reference point of the image (c) as compared with (b). The mismatch between the images results in a reduction of their common area, which is a limitation in the stereoscopic analysis.

In order to enhance the accuracy of the grid recognition, the noise of the calibration images can be reduced by averaging. Figure 5.4 shows a subdomain of a calibration image of camera 1 (left). The noise effect is visible in the background. Arithmetic averaging over 80 images reduces the noise effect (right). The intensity distribution along a row of pixels (shown inverted) reveals the reduction of noise after averaging.

The result of back-transformation using the method of Ch. 2 is shown in Fig. 5.5. The recombination of the dewarped calibration images shows their common region and a reduction in the effective region of the measurements. The deviation of the refined grid from the original grid in the object plane is shown in Fig. 5.6. Except in the vicinity of an imperfection in the calibration image, the difference between the grid points is under 0.4 pixel in both cameras. The distribution of the deviations is not homogeneous. This is due to the quality of the recognized grid, which is dependent on the accuracy of the grid in the object plane and the accuracy in the determination of its node locations in the image plane.

A comparison between the back-projection method presented in Ch. 2 and the reconstruction by an analytic projection is shown in Fig. 5.7. The projection function used, is the ratio of two second order two-variable polynomials, which is fitted to the grid points by the non-linear Levenberg-Marquardt algorithm. The images were processed by "PIVmap3", a module of the "PIVview" software[1], and were modified for better view. The expected locations of the grid points are marked with crosses. The discrete method of Ch. 2, Fig. 5.7a, shows the same quality of transformation for all nodes. The analytic projection, Fig. 5.7b, has a good quality of transformation in the inner region of the image, but shows considerable mismatch between the actual and the expected locations of the nodes near the boundaries.

In the image reconstruction by fitting analytical functions, the result of the transformation will have the characteristics of the mapping function. It can compensate for the presence of local imperfections in the calibration image by using smooth mapping functions. The effect of such imperfections is seen as local dislocations in Fig. 5.6. If the behavior of an analytic mapping function matches the real transformation function of an optical system, the stochastic effects such as image noise and the local dislocations of the grid points will be compensated during the reconstruction. This is similar to the behavior of the mean function of sample, which is defined in Sect 3.4.

The advantage of the reconstruction method presented in Ch. 2 is the automatic determination of the source functions, which in turn determines the transformation equations. The presence of dislocations in the identified grid causes departures of the source functions from their correct values. The improvement of calibration image, the enhancement of optical setup and camera settings for better image quality, and the improvement of node identification procedure are some methods for improving the accuracy of the transformation. One method for the improvement of the quality of the calibration image by averaging is shown in Fig. 5.4.

5.3 Measurements

Figure 5.8 shows typical records of both cameras. The stream of particles is made visible by the light-sheet. Visible machine parts are black painted to reduce light reflection. The illuminated particles, regions C in the figure, cover a considerable portion of the visible region. Secondary light reflections from the impeller leading edges, the impeller hub and the blade surfaces near the hub are strong enough to distort the particle images (or the *signal*). As the seeding particles attach to the exposed surfaces, an oil film is formed, which intensifies the effects of the secondary light reflections. A part of the signal

---

[1]PIVview, ver. 2.5, PivTec GmbH, Stauffenbergring 21, D-37075 Göttingen Germany. Homepage: http://www.pivtec.com

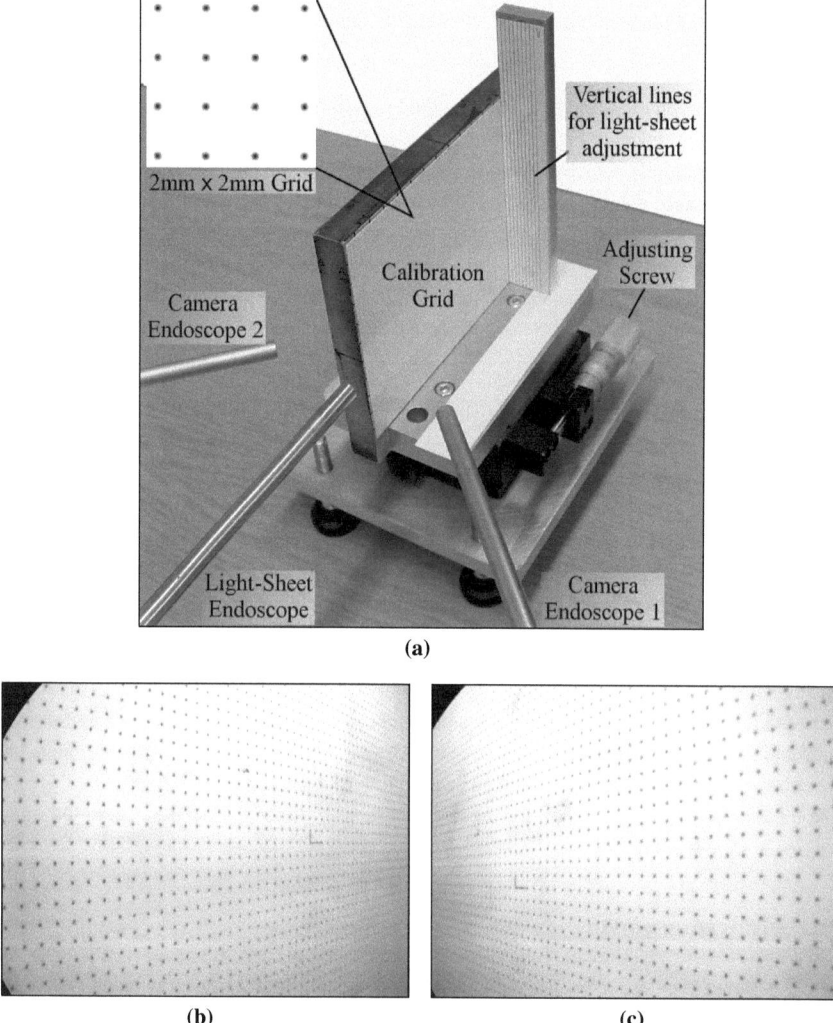

Fig. 5.3: (a) the calibration grid and its adjusting mechanism, (b) and (c) the calibration images of camera 1 and camera 2, respectively
The calibration plane can be adjusted in the horizontal plane with 0.01 mm accuracy. At each horizontal position, it can be rotated by relocation for the second camera.

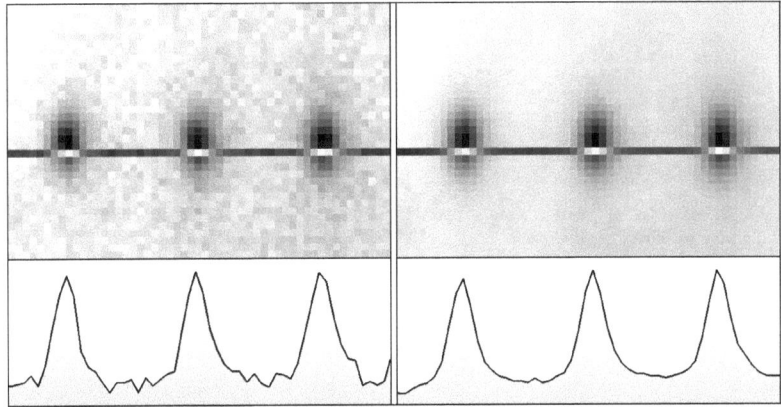

Fig. 5.4: A sub-domain of a calibration image of camera 1, before (left) and after (right) averaging
The image contrast and color values have been modified for better visibility. As a result of averaging, the noise level in the averaged image, visible in the background, is considerably reduced. The diagram shows the light intensity distribution along the inverted row of pixels with black as the minimum and white as the maximum intensities.

is lost due to the partial mismatch between the camera records as shown in Fig. 5.5c. The slightly out of focus regions of the images reduce the signal quality. These effects reduce the effective area of the light-sheet to a smaller measurable region as shown in Fig. 5.2. The regions, where the signal is distorted by light reflection differ between the cameras. Since the signal of both cameras are needed for the stereoscopic analysis, the lack of the signal of one camera makes the equivalent signal of the other useless. For this reason, the union of the regions of both cameras, within which the signal is lost, is subtracted from their common visible region.

The reconstructed images are shown in Figs. 5.8c and d. The image quality differs between the cameras. Camera 2 shows a partially out of focus region in the shroud-side, within which the quality of the cross-correlation is reduced.

Two further effects, which distort the PIV measurements, are shown in Fig. 5.9. The location, direction, and quality of the light-sheet is dependent on the position of the laser arm. Small movements can result in considerable changes in the light-sheet position. One source of movement is the vibration of the machine, which propagates both through the machine foundation and through the connection of the laser arm to the light-sheet endoscope. The light-sheet reflection, region A, which should be parallel with the marking B, is rotated due to the movement of the laser arm. This movement brings the

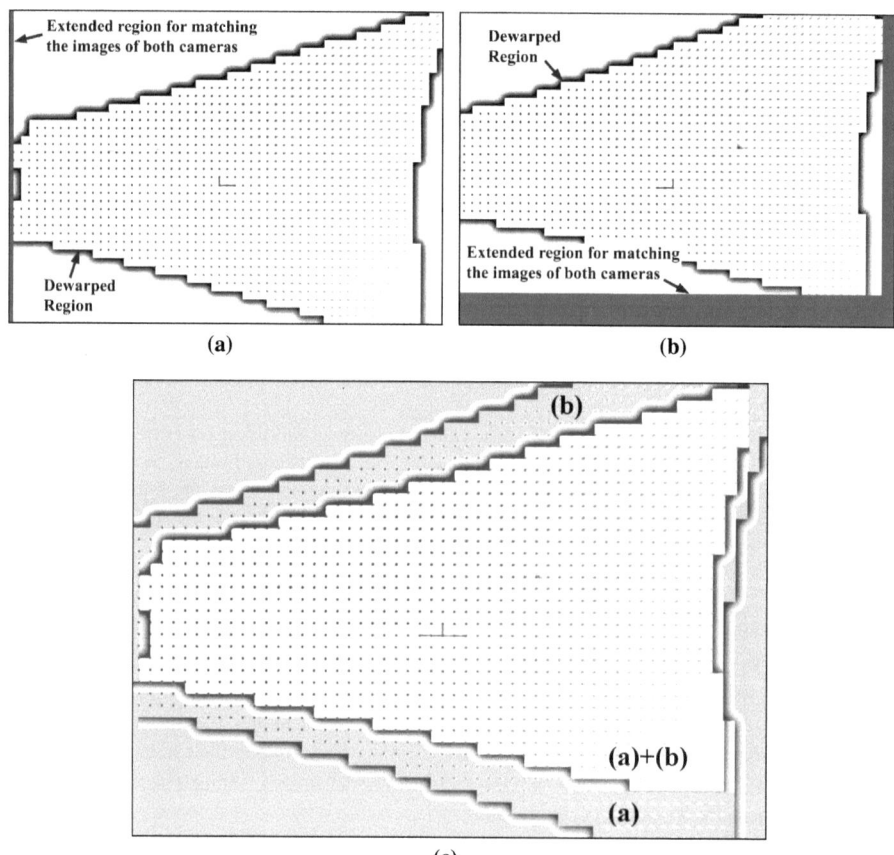

Fig. 5.5: The dewarped calibration images of (a) camera 1, (b) camera 2, and (c) the recombined dewarped images of cameras 1 and 2
The image of camera 2 is flipped around the vertical axis, in order to match the image of camera 1. The image areas are extended so that their reference points, the corner of the horizontal L in the grid, coincide with each other. The white region represents the common region of the cameras.

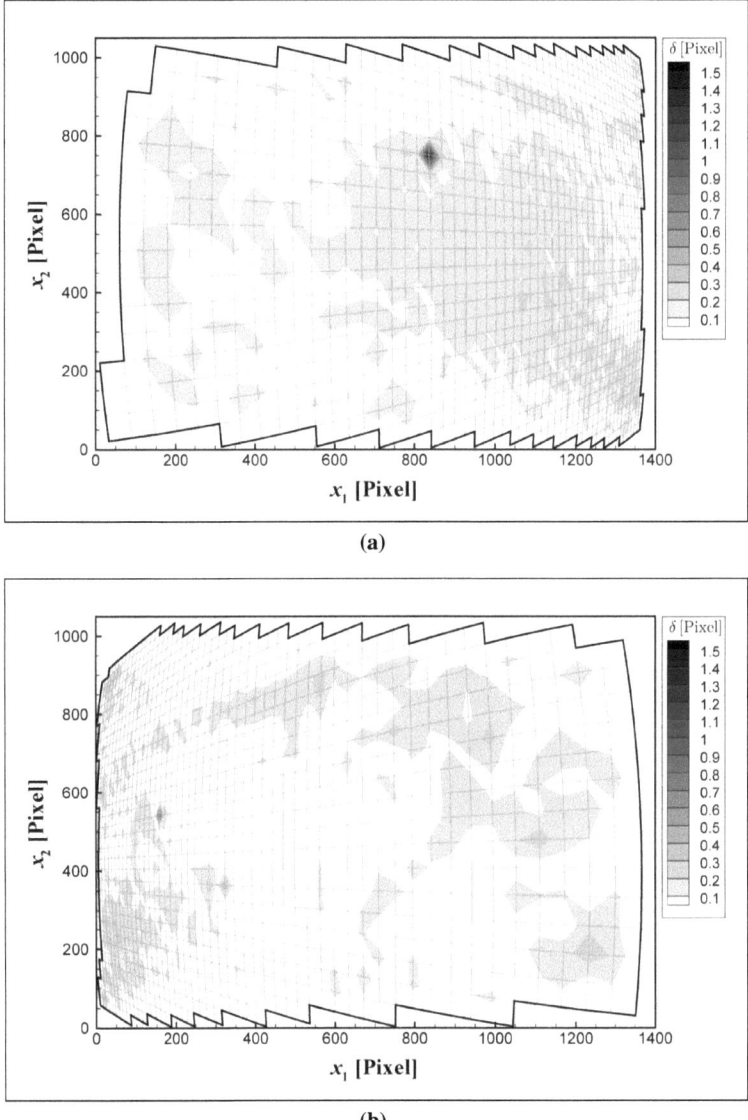

Fig. 5.6: The distribution of the distances between the nodes in the original grid and their corresponding nodes in the refined grid **(a)** camera 1 and **(b)** camera 2

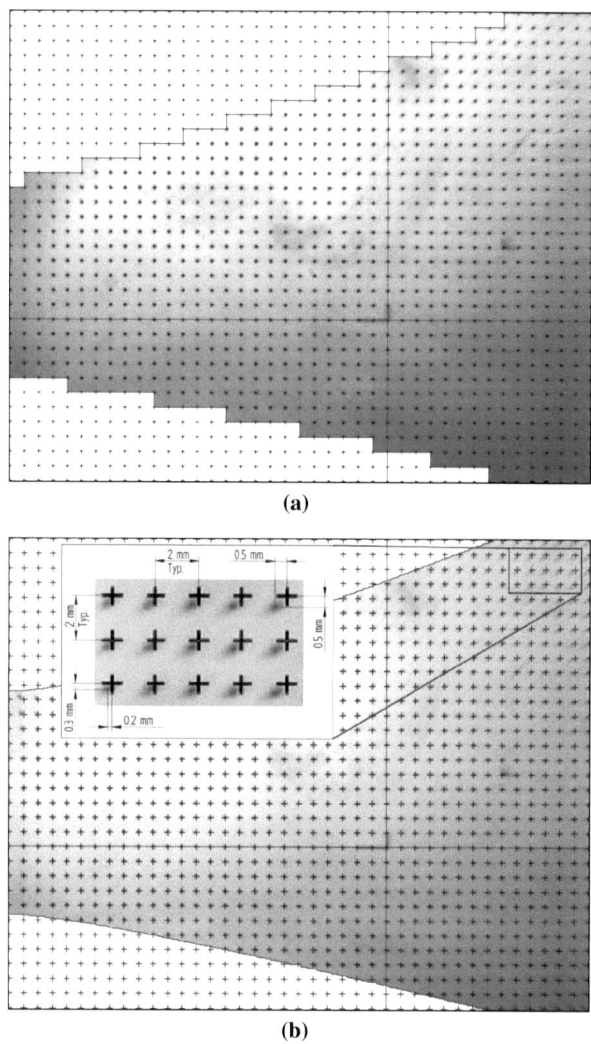

Fig. 5.7: Comparison between **(a)** the reconstruction method presented in Ch. 2 and **(b)** reconstruction by a second order analytic projection
The locations of the grid points are marked by the Gaussian patterns. The crosses show the expected locations of the grid points. The enlarged view shows the deviation of the grid points (the Gaussian patterns) from their expected locations in the border region of image (b).

light-sheet out of its calibrated position and is a source of error in the measurements. The out-of-plane components of the velocity field are affected by the light-sheet misalignment and are excluded from the results. The correction of the light-sheet misalignment requires the use of image processing besides the improvement of the measurement setup.

The accumulation of the seeding oil droplets on the optical path is another effect, which reduces or distorts the signal quality. Region C in Fig 5.9 shows blurred particle images due to droplet formation. This effect can be reduced by providing the seeding only during the measurements. However, after the third or fourth series of recordings, a cleaning of the optical ports may be necessary.

If the light-sheet matches the calibration plane, its reflection patterns in both images should coincide after image reconstruction. The movement of the light-sheet or its misalignment with respect to the calibration plane leads to the mismatch of these patterns as shown in Fig. 5.10. The exact alignment of the light-sheet in the calibration plane is also dependent on its light intensity distribution. In internal flow applications, where the measurement section is not directly accessible, the exact alignment of the light-sheet is hard to achieve. Partial deviation of the light-sheet from its calibration plane can be corrected during the data analysis (Coudert and Schon, 2001). A method for the correction of the light-sheet position has not been included in this work and is left as its future extension.

The applicability of the SPIV and the effect of the different sources of error presented in this section for the flow investigation at the inlet of the compressor is further studied by the analysis of one typical set of measurements in the next section.

5.4 Data analysis and results

In this section the analysis of one typical set of SPIV records is presented and the effect of several sources of error, discussed in the previous section, are further studied by quantitative results.

The set of PIV records are visually checked for quality. Acceptable images are dewarped by using the reconstruction data from the calibration images. The dewarped images are then analyzed by the "PIVview" software. The following results are the averaged values over the velocity fields of each pair of the records. The operating conditions of the compressor is the same as in the pressure probe measurements presented in Sect. 4.6.

Figure 5.11 shows the steady state velocity distribution obtained by averaging in the light-sheet plane. As expected from the flow condition at the zero setting of the IGV, the velocity field is uniform. The velocity distributions on the cross-section of two planes, plane 1 and plane 2, with the light-sheet plane have been selected for comparison with the pressure probe results. The location of these planes are 15.6 mm and 19.4 mm downstream of the pressure probe plane 3, Fig. 5.2, respectively. The obviously incorrect velocity vectors have been filtered out. However, the magnitude of some velocity vectors in

104 Chapter 5. PIV Measurement at the Compressor Inlet

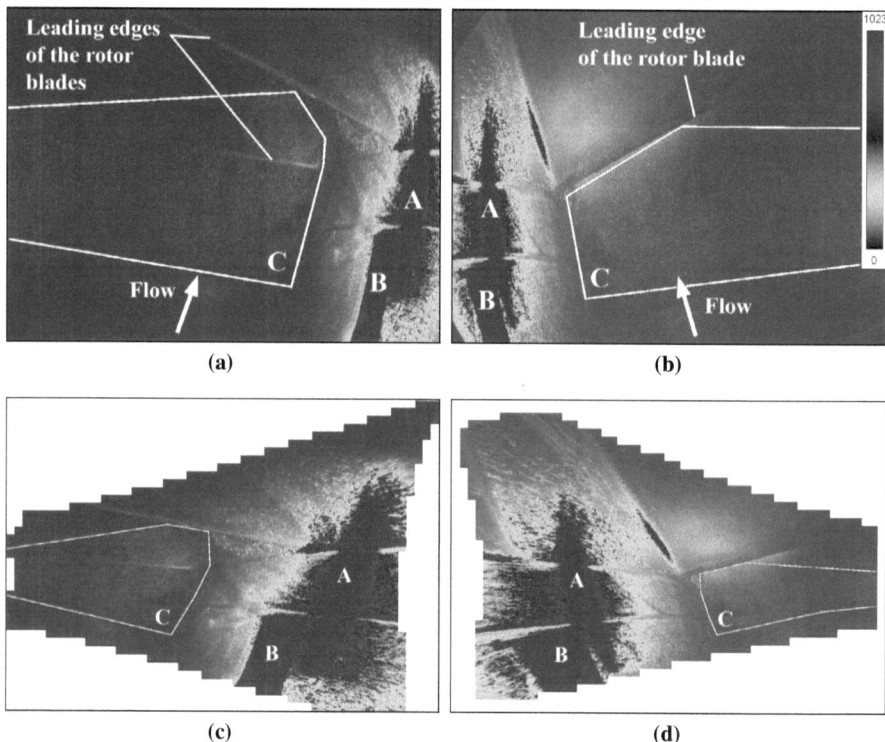

Fig. 5.8: Typical PIV records (**a**) camera 1, (**b**) camera 2, (**c**) camera 1 dewarped, and (**d**) camera 2 dewarped
Region A is the reflection of the light-sheet and region B is a rectangular marking on the rotor for the adjustment of the light-sheet. The light-sheet should be parallel to this rectangular region. Region C shows the measurable part of the image, where the illuminated particles are distinct from the background. The original gray-scale levels are replaced by the levels shown in the color bar of image (b) for better visibility. The color bar is valid for all images.

Chapter 5. PIV Measurement at the Compressor Inlet 105

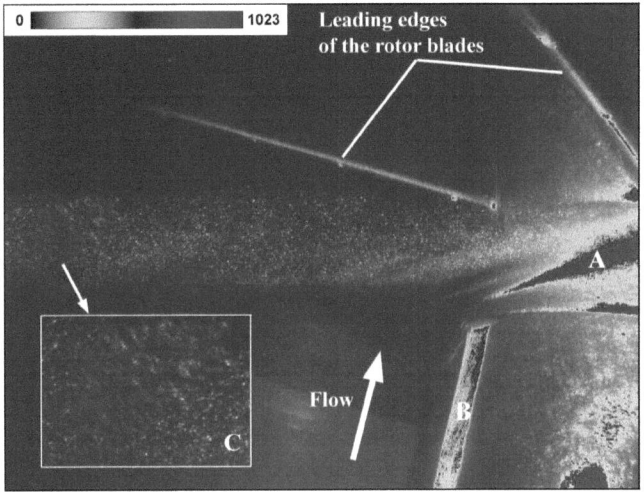

Fig. 5.9: The rotation of the light-sheet, region A, from its expected direction, which is along the machine axis and parallel to the rectangle B, and the effect of droplet formation on the end window of the endoscope, region C
The original gray-scale levels are replaced by the levels shown in the color bar for better visibility.

the upper right and the lower left regions of the vector field is physically not justifiable. The variation of the magnitude from the inner field toward these regions is continuous, which makes the verification of the vector values difficult. A comparison of the velocity measurements by the pressure probes and PIV is presented in Fig. 5.12. The PIV velocity values are the results of averaging over 55 double-images. The axial velocity matches the pressure probe measurement result within its uncertainty range. The radial component, V_2, shows more departure from the pressure probe result than the axial component, V_1. For smoother velocity curves, the number of the PIV records should be increased.

The axial and radial velocity components measured by PIV show good correspondence with the pressure probe result in the range shown in Fig. 5.12. The out-of-plane or the circumferential component of the velocity, V_3, is highly affected by the light-sheet misalignment and is excluded from this diagram. Out of the shown range, the PIV velocity components show departure from the probe measurements. This is noticed as a change in the magnitude of the velocity vectors in the top right and the bottom left regions of Fig. 5.11.

Some of the effects, which reduce the measurable region of the PIV images, were presented in the previous section. The misalignment of the light-sheet is a source of error, which affects the accuracy of

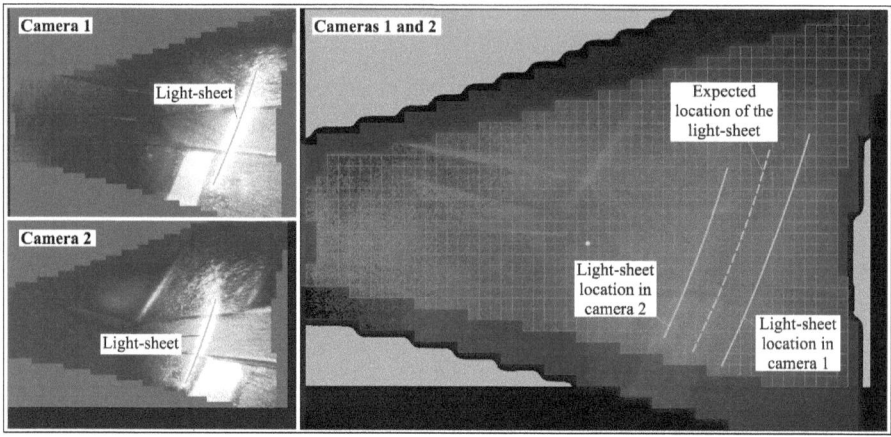

Fig. 5.10: The location of the light-sheet reflection patterns after the image reconstruction for the case of misalignment between the light-sheet and the calibration plane. The approximate expected location of the light-sheet is shown by a dashed line. The grid spacing is 2 mm × 2 mm.

the measurements and can cause gradual change in the velocity magnitude. Machine vibration and the requirement of the optical adjustment of the light-sheet during the measurements are the main causes of the light-sheet misalignment.

The relative movement of optical setup due to the machine vibration is another source of error in the measurements. It is an stochastic phenomenon and is supposed to be suppressed by averaging. However, its limits, within which the measured values are valid, need to be determined.

The formation of the oil film, from the seeding particles, on the optical paths is a source of distortion in the PIV images. This effect is shown in the region C of Fig. 5.9. Since the formation of the oil film is gradual, the identification or the definition of a threshold for its effect is not easy. One method of the verification of the resultant velocity field and the determination of its acceptable range is the uncertainty analysis of the measurement procedure. This analysis is left as a future extension of the measurement methods presented in this work.

Chapter 5. PIV Measurement at the Compressor Inlet   107

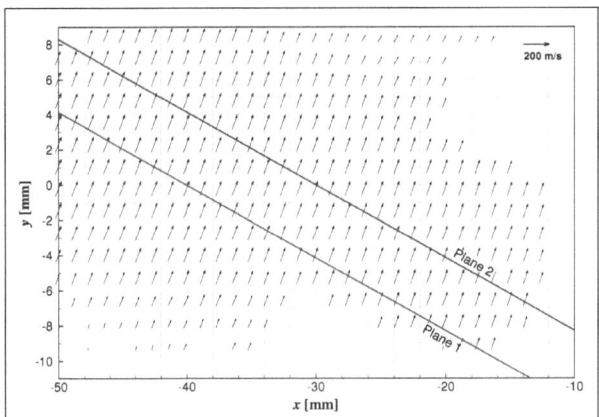

Fig. 5.11: The projected average velocity field on the light-sheet plane
The coordinates xy and the location of the planes 1 and 2 in the flow channel are shown in Fig. 5.2.

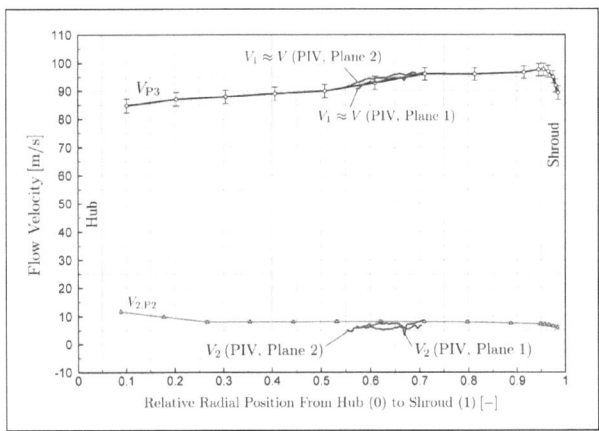

Fig. 5.12: Comparison between the pressure probe and the PIV measurements
Shown are the velocity magnitude, V, and its axial, V_1, and radial, V_2, components, see Fig. 4.8. The circumferential component of the velocity, V_3, is highly affected by the light-sheet misalignment and is excluded from the diagram. The indices P2 and P3 indicate the pressure probe planes 2 and 3, respectively. The PIV planes 1 and 2 are defined in Figs. 5.2 and 5.11.

Chapter 6

Conclusions and Future Work

The main goal of this research work is the development of the endoscopic stereoscopic PIV for the flow investigation in the inlet of a centrifugal compressor. An endoscopic SPIV setup has been designed and constructed, in which several limitations in the application of this measurement technique in the internal flow investigations have been solved so that the achievable quality of the PIV records is comparable to that of the traditional PIV.

An existing pressure probe flow measurement system is used in order to verify the PIV results. In order to improve the accuracy of the flow measurements by the pressure probes, a theory for calibration under quasi-steady conditions is presented in Ch. 3. Compared to the current methods of calibration, in which the mean values are considered constant, in this theory the time variation of the mean values is considered in the calibration procedure. It provides the uncertainty analysis of the quasi-steady data. Starting from the calibration data, this chapter provides the mathematical basis for a complete analysis of the calibration and measurement data and provides methods for the estimation of the measurement error. By the inclusion of the external and internal influences on the calibration data, the theory not only improves the accuracy of the calibration, but also provides a method for the on-site calibration, where the stable conditions are hard to achieve.

The statistical analysis presented is based on the direct use of the calibration records without the implementation of the statistical distribution functions. The theory can be extended to include the distribution functions and the methods of finding the matching distribution function to a series of measurements. Also it can be extended to include the multi-point – multi-point mode of data acquisition, which has not been covered in Ch. 3.

Endoscopic SPIV has been used for the first time to investigate the inlet flow in a centrifugal compressor. Several practical features of the endoscopic measurements have been discussed in Ch. 5. The results show the sensitivity of the endoscopic stereoscopic measurements of the internal flows to the reflection of the light from the visible machine components, the quality of the optics, the stability of the position of the light-sheet, the formation of oil film and droplets on the optical paths, the reconstruction of the recordings, the alignment of the light-sheet and calibration plane, and the relative movement of the optical components due to the machine vibration. The measurements show the capability of the measurement setup to measure the velocity field. However, an uncertainty analysis is required in order to verify the results and to determine the extent of the valid data.

In internal flow measurements, the fine adjustment of the light-sheet is hard to achieve. A data processing method is required for the fine adjustment of the light-sheet on the calibration plane. Addition of a mechanism, which provides movements for the cameras and the laser endoscope from the measurement position to the cleaning and calibration positions, is another improvement, which facilitates the long-term measurements.

The reconstruction of the distorted endoscopic images has been considered in detail in Ch. 2. A novel method for distortion compensation based on partial differential equations has been presented. Compared to the current analytical methods, which are capable of partial reconstruction of the endoscopic distortions, the presented method is capable of full reconstruction of the distorted images. The method provides a unique solution to a distortion pattern and is independent from the user interference or judgment. Several examples presented in Chs. 2 and 5 show the capability of the method to reconstruct highly distorted images.

The method is based on the iterative numerical solution of the discretized Poisson equations. The solution of the transformation functions takes considerably more time than the analytical mappings. Therefore, the method is not suitable for real-time data analysis. The optimization of the solution procedure or the implementation of a new solution algorithm to reduce the time of the transformation can be considered as a future work. The implementation of partial differential equations as transformation functions provides a new class of reconstruction methods for further improvements of the PIV measurement technique and other image-based optical measurement methods.

Bibliography

Baggenstoss, P. M. Image distortion analysis using polynomial series expansion. *IEEE Trans. Pattern Anal. Mach. Intell.*, 26:1438–1451, 2004.

Beckwith, T. G. and Marangoni, R. D. *Mechanical measurements*. Addison Wesley, 4th ed., 1990. ISBN: 0-201-17866-4.

Bronshtein, I. N., Muehlig, H., Musiol, G., and Semendyayev, K. A. *Handbook of Mathematics*. Springer, 5th ed., 2007. ISBN: 3-540-72121-5.

Bubolz, T. *Untersuchungen von randzonenkorrigierten Axialverdichterbeschaufelungen mit Strömungs-Meßsonden*. PhD thesis, Fachbereich Maschinenbau der Universität Hannover, Hannover, Germany, 2005.

Coudert, S. J. M. and Schon, J.-P. Back-projection algorithm with misalignment corrections for 2D3C stereoscopic PIV. *Meas. Sci. Technol.*, 12:1371–1381, 2001.

Dierksheide, U., Meyer, P., Hovestadt, T., and Hentschel, W. Endoscopic 2D particle image velocimetry (PIV) flow field measurements in IC engines. *Exp. Fluids*, 33:794–800, 2002. doi: 10.1007/s00348-002-0499-3.

Dillmann, A., Loose, S., Raffel, M., and Meier, G. E. A. Strömungsmess- und Vesuchstechnik. Vorlesungsskript, Deutsches Zentrum für Luft- und Raumfahrt (DLR), Göttingen, 2007.

Gizzi, W. P. *Dynamische Korrekturen für schnelle Strömungssonden in hochfrequent fluktuierenden Strömungen*. PhD thesis, Eidgenössische Tech. Hochsch. Zürich, 2000. Nr 13482.

Hagelstein, D., Prinsier, J., and Van den Braembussche, R. A. Vorleitradoptimierung I: Vorleitrad - Laufrad Strömungsoptimierung und Interaktion – Abschlussbericht: Vorhaben Nr 705. Technical Report 726, Forschungsvereinigung Verbrennungskraftmaschinen e.V. (FVV), 2001.

Hewlett-Packard. HP 34970A data acquisition / switch unit, user's guide: Part number 34970-90002. Hewlett-Packard Co., 1997.

Kalibrierschein DPI 610. Kalibrierschein für das elektrisches Druckmessgerät DPI 610: Nr 02-10-05 DKD-K-40601 2005-02. Kalibrierlaboratorium für die Messgrösse Druck, Deutscher Kalibrierdienst (DKD), 2005.

Kegalj, M. and Schiffer, H.-P. Endoscopic PIV measurements in a low pressure turbine rig. *Exp. Fluids*, 47:689–705, 2009. doi: 10.1007/s00348-009-0712-8.

Kincaid, D. R., Cheney, E. W., and Cheney, W. *Numerical analysis: Mathematics of scientific computing*. Brooks/Cole, 2nd print, 1991. ISBN: 0-534-13014-3.

Levenberg, K. A method for the solution of certain non-linear problems in least-squares. *Q. Appl. Math.*, 2(2):164–168, 1944.

Liu, B., Yu, X., Liu, H., Jiang, H., Yuan, H., and Xu, Y. Application of SPIV in turbomachinery. *Exp. Fluids*, 40, 2006. doi: 10.1007/s00348-005-0102-9.

Marquardt, D. W. An algorithm for least-squares estimation of nonlinear parameters. *J. Soc. Ind. Appl. Math.*, 11(2):431–441, 1963.

Michalski, L., Eckersdorf, K., and McGhee, J. *Temperature measurement*. Wiley, 2nd ed., 2001. ISBN: 0-471-86779-9.

Mohseni, A., Goldhahn, E., Van den Braembussche, R. A., and Seume, J. R. Novel IGV designs for centrifugal compressors and their interaction with the impeller. *Proc. of the ASME Turbo Expo 2010: Power for Land, Sea and Air, June 14-18, Glasgow, UK*, 2010. No. GT2010-23048.

Mohseni, A. Solution of two-dimensional Navier-Stokes equations using multi-block scheme (in Persian). Master's thesis, Department of Mechanical Engineering, Faculty of Engineering, University of Tehran, Tehran, Iran, 2000.

Natterer, F. Imaging and inverse problems of partial differential equations. *Jahresber. Deutsch. Math.*, 109:31–48, 2006.

Pollak, C., Stubbings, T., and Hutter, H. Differential image distortion correction. *Microsc. Microanal.*, 7:335–340, 2001. doi: 10.1007/s10005-001-0007-1.

Raffel, M., Kompenhans, J., Wereley, S. T., and Willert, C. E. *Particle Image Velocimetry: A Practical Guide*. Exp. Fluid Mech. Springer, 2nd ed., 2007. ISBN: 978-3-540-72307-3.

Scheimpflug, T. Improved method and apparatus for the systematic alteration or distortion of plane pictures and images by means of lenses and mirrors for photography and for other purposes. British Patent, 1904.

Seume, J. R., Goldhahn, E., Steglich, T., Van den Braembussche, R. A., Prinsier, J., Alsalihi, Z., and Verstraete, T. Vorleitradoptimierung II: Strömungsvorgänge zwischen Vorleitapparat und Laufrad eines Radialverdichters und ihre gegenseitige Beeinflussung - Abschlussbericht: Vorhaben Nr. 846. Technical Report 838, Forschungsvereinigung Verbrennungskraftmaschinen e.V. (FVV), 2007.

Soloff, S. M., Adrian, R. J., and Liu, Z.-C. Distortion compensation for generalized stereoscopic particle image velocimetry. *Meas. Sci. Technol.*, 8:1441–1454, 1997.

Steger, J. L. and Sorenson, R. L. Automatic mesh-point clustering near a boundary in grid generation with elliptic partial differential equations. *J. Comput. Phys.*, 33:405–410, 1979.

Tai, X.-C., Chan, T. F., Lie, K.-A., and Osher, S., editors. *Image Processing Based on Partial Differential Equations: Proceedings of the International Conference on PDE-Based Image Processing and Related Inverse Problems, CMA, Oslo, August 8–12, 2005*. Math. and Vis. Springer, 2007. ISBN: 3-540-33266-9. doi: 10.1007/978-3-540-33267-1.

Tannehill, J. C., Anderson, D. A., and Pletcher, R. H. *Computational fluid mechanics and heat transfer*. Ser. in comput. and phys. process. in mech. and therm. sci. Taylor & Francis, 2nd ed., 1997. ISBN: 1-560-32046-X.

Tropea, C., Foss, J. F., and Yarin, A. L. *Springer Handbook of Experimental fluid mechanics*. Springer, 2007. ISBN: 978-3-540-25141-5.

Udrea, D. D., Bryanston-Cross, P. J., Moroni, M., and Querzoli, G. Particle tracking velocimetry techniques. In Stanislas, M., Kompenhans, J., and Westerweel, J., editors, *Particle Image Velocimetry: Progress towards industrial application*, volume 56 of *Fluid mech. and its appl.*, pages 279–304. Kluwer Acad. Press, 2000. ISBN: 0-792-36160-1.

Van den Braembussche, R. A., Prinsier, J., and Doulgeris, G. Design and performance evaluation of 3 different inlet guide vanes for radial compressors. Technical report, von Kármán Inst. for Fluid Dyn. (VKI), 2006. No. 2007-03.

Voges, M., Beversdorff, M., Willert, C., and Krain, H. Application of particle image velocimetry to a transonic centrifugal compressor. *Exp. Fluids*, 43:371–384, 2007. doi: 10.1007/s00348-007-0279-1.

Weisstein, E. W. *CRC concise encyclopedia of mathematics*. Chapman & Hall/CRC, 2nd ed., 2003. ISBN: 1-584-88347-2.

Wernet, M. P., Bright, M. M., and Skoch, G. J. An investigation of surge in a high-speed centrifugal compressor using digital PIV. *ASME J. Turbomach.*, 123:418–428, 2001. doi: 10.1115/1.1343465.

Westerweel, J. *Digital Particle Image Velocimetry: Theory and Application*. Delft Univ. Press, 1993. ISBN: 9-062-75881-9.

Wilson, D. G. and Korakianitis, T. *The design of high-efficiency turbomachinery and gas turbines*. Prentice Hall, 2nd ed., 1998. ISBN: 0-133-12000-7.

Woisetschläger, J. and Göttlich, E. Recent applications of particle image velocimetry to flow research in thermal turbomachinery. In Schroeder, A. and Willert, C. E., editors, *Particle Image Velocimetry*, volume 112 of *Top. in Appl. Phys.* Springer, 2008. ISBN: 978-3-540-73527-4.

Yun, Y. I., Porreca, L., Kalfas, A. I., Song, S. J., and Abhari, R. S. Investigation of three-dimensional unsteady flows in a two-stage shrouded axial turbine using stereoscopic PIV – kinematics of shroud cavity flow. *ASME J. Turbomach.*, 130:011021, 2008. doi: 10.1115/1.2720873.

Zachau, U., Buescher, C., Niehuis, R., Hoenen, H., Wisler, D. C., and Moussa, Z. M. Experimental investigation of a centrifugal compressor stage with focus on the flow in the pipe diffuser supported by particle image velocimetry (PIV) measurements. *Proc. of the ASME Turbo Expo 2008: Power for Land, Sea and Air, GT2008, June 9–13, Berlin, Germany*, 2008. No. GT2008-51538.

Appendix A

Polynomial Fit with the Method of Least Squares

Considering a sample $\{(t_i, x_i)\}$, $i \in [1, N]_{\mathbb{N}}$, in the Cartesian coordinates tx, and a real function $x(t; a_k)$ with independent variable $t \in \mathbb{R}$ and parameters $a_k \in \mathbb{R}$, the least squares method is a mathematical procedure for finding a set of parameters a_k, so that the sum of the squares of the offsets or the residuals of the sample points is minimized (Weisstein, 2003).

Considering a polynomial of order P:

$$x(t; a_k) = \sum_{k=0}^{P} \left(a_k \, t^k \right) \tag{A.1}$$

and the sum of the squares of the vertical offsets:

$$R(a_k) = \sum_{i=1}^{N} [x_i - x(t_i; a_k)]^2 \tag{A.2}$$

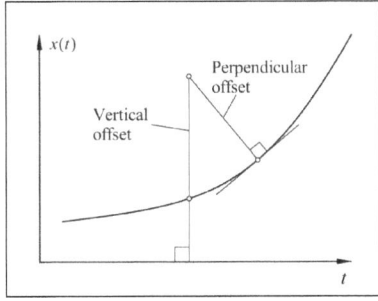

Fig. A.1: The vertical and perpendicular offsets of a point from a curve

for $R(a_k)$ to be a minimum, it is required that:

$$\frac{\partial R(a_k)}{\partial a_k} \equiv R_{,k}(a_k) = 0, \quad \forall k \in [0, P]_\mathbb{I} \qquad (A.3)$$

Inserting from Eqs. (A.1) and (A.2) into Eq. (A.3) results in:

$$\frac{\partial}{\partial a_k} \sum_{i=1}^{N} \left[x_i - \sum_{k'=0}^{P} \left(a_{k'} \, t_i^{k'} \right) \right]^2 = 0$$

$$\therefore \quad \sum_{i=1}^{N} \left[t_i^k \left(x_i - \sum_{k'=0}^{P} \left(a_{k'} \, t_i^{k'} \right) \right) \right] = 0, \quad \forall k \in [0, P]_\mathbb{I} \qquad (A.4)$$

$$\therefore \quad \sum_{k'=0}^{P} \left[a_{k'} \sum_{i=1}^{N} \left(t_i^{k'} t_i^{k} \right) \right] = \sum_{i=1}^{N} \left(x_i \, t_i^{k} \right), \quad \forall k \in [0, P]_\mathbb{I} \qquad (A.5)$$

Using the matrix notation, Eq. (A.5) can be written as:

$$\boldsymbol{A} \cdot \boldsymbol{X} = \boldsymbol{B} \qquad (A.6)$$

where the matrices $\boldsymbol{A}$, $\boldsymbol{X}$ and $\boldsymbol{B}$ are as follows:

$$\begin{aligned}
\boldsymbol{A} &= [A_{ij}]_{(P+1)\times(P+1)}, & A_{ij} &= \sum_{k=1}^{N} \left(t_k^i \, t_k^j \right) \\
\boldsymbol{X} &= [X_i]_{(P+1)\times 1}, & X_i &= a_i \\
\boldsymbol{B} &= [B_i]_{(P+1)\times 1}, & B_i &= \sum_{k=1}^{N} \left(x_k \, t_k^i \right)
\end{aligned} \qquad (A.7)$$

The solution of Eq. (A.7) is a set of coefficients a_k, which determines a polynomial $x(t; a_k)$, Eq. (A.1), with an order not grater than P.

For $k = 0$, Eq. (A.4) shows that for a least squares fit, the sum of the offsets is equal to zero. Therefore:

Theorem A.1 Let a polynomial $x(t; a_k) = \sum_{k=0}^{P} \left(a_k \, t^k \right)$ be a least squares fit based on vertical offsets to a sample $\{(t_i, x_i)\}$, $i \in [1, N]_\mathbb{N}$. Then the sum of the vertical offsets over all points will vanish:

$$\sum_{k=0}^{P} [x_i - x(t_i; a_k)] = 0 \qquad (A.8)$$

□

Appendix B

Spline Interpolation

A *spline function of degree* k having the *knots* t_i, $i \in [0,\ N]_{\mathbb{I}}$ and $t_i < t_{i+1}$, is a function $S(t) : \mathbb{R} \longmapsto \mathbb{R}$ such that (Kincaid et al., 1991):

1. On each interval $[t_i,\ t_{i+1})$, $S(t)$ is a polynomial of a degree not grater than k.
2. $S(t)$ is C^{k-1} continuous on $[t_0,\ t_N]$

For a given set of points $\{(t_i,\ x_i)\}$, $i \in [0,\ N]_{\mathbb{I}}$ and $t_i < t_{i+1}$, a spline function of order three, also known as *cubic spline*, is a function $S(t)$:

$$S(t) = \begin{cases} S_0(t) & t \in [t_0,\ t_1] \\ S_1(t) & t \in [t_1,\ t_2] \\ \vdots & \vdots \\ S_{N-1}(t) & t \in [t_{N-1},\ t_N] \end{cases} \qquad (B.1)$$

in which $S_i(t) = a_{0,i} + a_{1,i}\, t + a_{2,i}\, t^2 + a_{3,i}\, t^3$, $i \in [0,\ N-1]_{\mathbb{I}}$, are cubic polynomials and satisfy:

$$\begin{aligned} x_0 &= S_0(t_0) \\ x_N &= S_{N-1}(t_N) \\ x_j &= S_{i-1}(t_i) = S_i(t_i)\,, \qquad i \in [0,\ N-1]_{\mathbb{I}} \\ S'_{i-1}(t_i) &= S'_i(t_i) \\ S''_{i-1}(t_i) &= S''_i(t_i) \end{aligned} \qquad (B.2)$$

where $S'(t) := dS/dt$ and $S''(t) := d^2S/dt^2$. Defining $z_i := S''_i(t_i)$ and $h_i := t_{i+1} - t_i$, the second derivative of $S_i(t)$ is calculated as:

$$S''_i(t) = \frac{1}{h_i}\left[z_i(t_{i+1} - t) + (t - t_i)\right]\,, \qquad i \in [0,\ N-1]_{\mathbb{I}} \qquad (B.3)$$

Integrating Eq. (B.3) using conditions (B.2) results in:

$$S_i(t) = \frac{1}{6h_i}\left[z_i(t_{i+1}-t)^3 + z_{i+1}(t-t_i)^3\right]$$
$$+ \left(\frac{x_{i+1}}{h_i} - \frac{z_{i+1}h_i}{6}\right)(t-t_i) + \left(\frac{x_i}{h_i} - \frac{z_i h_i}{6}\right)(t_{i+1}-t), \quad i \in [0, N-1]_\mathbb{I} \qquad (B.4)$$

$$S'_i(t) = \frac{1}{2h_i}\left[-z_i(t_{i+1}-t)^2 + z_{i+1}(t-t_i)^2\right]$$
$$+ \frac{x_{i+1}-x_i}{h_i} - \frac{h_i}{6}(z_{i+1}-z_i), \quad i \in [0, N-1]_\mathbb{I} \qquad (B.5)$$

The continuity condition for $S'(t)$, Eq. (B.2), is used to determine z_i:

$$h_{i-1}z_{i-1} + 2(h_{i-1}+h_i)z_i + h_i z_{i+1} = 6\left(\frac{x_{i+1}-x_i}{h_i} - \frac{x_i-x_{i-1}}{h_{i-1}}\right), \quad i \in [0, N-1]_\mathbb{I} \quad (B.6)$$

The solution of this tridiagonal diagonally dominant system of equations, requires that z_0 and z_N be known. In the case of $z_0 = z_N = 0$, $S(t)$ in Eq. (B.1) is called a *natural cubic spline*. It is evident that for a cubic spline fit, a minimum of three points is required, $N \geq 2$.

Appendix C

Gaussian Function

In a right handed Cartesian coordinate system $Oxyz$ in $\mathbb{R}^3$, the *elliptic Gaussian function* centered on the line (x_0, y_0, z) and rotated around it by an angle θ (measured from x^+-axis and positive in the direction toward y^+-axis) is (Weisstein, 2003):

$$z(x, y) = k \exp\left[-\frac{[(x - x_0)\cos\theta + (y - y_0)\sin\theta]^2}{a^2} - \frac{[-(x - x_0)\sin\theta + (y - y_0)\cos\theta]^2}{b^2} \right] \tag{C.1}$$

where $\{a, b, k\} \subset \mathbb{R}^+$ and $\theta \in (-\pi, \pi]$. An equivalent form of Eq. (C.1) is:

$$z(x, y) = \exp\left(c_{00} + c_{10}\, x + c_{20}\, x^2 + c_{01}\, y + c_{11}\, x\, y + c_{02}\, y^2 \right) \tag{C.2}$$

in which:

$$c_{00} = \ln(k) - \frac{(x_0 \cos\theta + y_0 \sin\theta)^2}{a^2} - \frac{(x_0 \sin\theta - y_0 \cos\theta)^2}{b^2} \tag{C.3a}$$

$$c_{10} = \frac{2\cos\theta\,(x_0 \cos\theta + y_0 \sin\theta)}{a^2} + \frac{2\sin\theta\,(x_0 \sin\theta - y_0 \cos\theta)}{b^2} \tag{C.3b}$$

$$c_{01} = \frac{2\sin\theta\,(x_0 \cos\theta + y_0 \sin\theta)}{a^2} - \frac{2\cos\theta\,(x_0 \sin\theta - y_0 \cos\theta)}{b^2} \tag{C.3c}$$

$$c_{11} = -2\sin\theta \cos\theta \left(\frac{1}{a^2} - \frac{1}{b^2} \right) \tag{C.3d}$$

$$c_{20} = -\frac{(\cos\theta)^2}{a^2} - \frac{(\sin\theta)^2}{b^2} \quad \Rightarrow \quad c_{20} \in \mathbb{R}^- \tag{C.3e}$$

$$c_{02} = -\frac{(\sin\theta)^2}{a^2} - \frac{(\cos\theta)^2}{b^2} \quad \Rightarrow \quad c_{02} \in \mathbb{R}^- \tag{C.3f}$$

or after rearrangement for 2θ:

$$c_{00} = \ln(k) - p(x_0^2 + y_0^2) - q\left[(x_0^2 - y_0^2)\cos(2\theta) + 2\,x_0\,y_0\sin(2\theta)\right] \tag{C.4a}$$

$$c_{10} = 2\,p\,x_0 + 2\,q\,(x_0\cos(2\theta) + y_0\sin(2\theta)) \tag{C.4b}$$

$$c_{01} = 2\,p\,y_0 + 2\,q\,(x_0\sin(2\theta) - y_0\cos(2\theta)) \tag{C.4c}$$

$$c_{11} = -2\,q\sin(2\theta) \tag{C.4d}$$

$$c_{20} = -p - q\cos(2\theta) \quad \Rightarrow \quad c_{20} \in \mathbb{R}^- \tag{C.4e}$$

$$c_{02} = -p + q\cos(2\theta) \quad \Rightarrow \quad c_{02} \in \mathbb{R}^- \tag{C.4f}$$

where:

$$p = \frac{1}{2}\left(\frac{1}{a^2} + \frac{1}{b^2}\right), \quad q = \frac{1}{2}\left(\frac{1}{a^2} - \frac{1}{b^2}\right), \quad p \in \mathbb{R}^+,\ q \in \mathbb{R} \tag{C.5}$$

$$a^2 = \frac{1}{p+q}, \qquad b^2 = \frac{1}{p-q} \tag{C.6}$$

The parameters x_0, y_0, θ, k, p, and q are calculated from Eqs. (C.4a) to (C.4f) as follows:

$$x_0 = \frac{2\,c_{10}\,c_{02} - c_{11}\,c_{01}}{\Delta} \tag{C.7a}$$

$$y_0 = \frac{2\,c_{20}\,c_{01} - c_{11}\,c_{10}}{\Delta} \tag{C.7b}$$

$$\tan(2\theta) = \frac{c_{11}}{c_{20} - c_{02}} \tag{C.7c}$$

$$k = \exp\left(c_{00} + \frac{c_{10}^2\,c_{02} - c_{10}\,c_{11}\,c_{01} + c_{20}\,c_{01}^2}{\Delta}\right) \tag{C.7d}$$

$$p = -\frac{c_{20} + c_{02}}{2} \tag{C.7e}$$

$$q = -\frac{c_{20} - c_{02}}{2\cos(2\theta)} \tag{C.7f}$$

where $c_{20} \neq c_{02}$, $\theta \neq \pm\pi/4$ and

$$\Delta := c_{11}^2 - 4\,c_{20}\,c_{02} \tag{C.8a}$$

Inserting from Eqs. (C.4d), (C.4e), (C.4f) in Eq. (C.8a) and using Eq. (C.6) gives:

$$\Delta = 4\,q^2\sin^2(2\theta) - 4\left(p^2 - q^2\cos^2(2\theta)\right)$$

$$\therefore \quad \Delta = 4\left(q^2 - p^2\right) = \frac{-4}{a^2\,b^2} \quad \Rightarrow \quad \Delta \in \mathbb{R}^- \tag{C.8b}$$

Also from Eqs. (C.4a) to (C.4f) the domain of θ is $(-\pi/2,\ \pi/2]$.

Appendix C. Gaussian Function

Special cases

Case I

When $c_{20} = c_{02}$, from Eqs. (C.4e) and (C.4f):

$$q \cos(2\theta) = 0 \Rightarrow \begin{cases} q = 0 \Leftrightarrow a = b \\ \text{or} \\ \theta = \pm \pi/4 \ (\text{see Case II}) \end{cases} \qquad \text{(C.9)}$$

With $q = 0$ or $a = b$ the Gaussian function is called a *circular Gaussian function* and is centered on the line (x_0, y_0, z). For this case Eqs. (C.4a) to (C.4f) are simplified as follows:

$$c_{00} = \ln(k) - p(x_0^2 + y_0^2) \qquad \text{(C.10a)}$$
$$c_{10} = 2\,p\,x_0 \qquad \text{(C.10b)}$$
$$c_{01} = 2\,p\,y_0 \qquad \text{(C.10c)}$$
$$c_{11} = 0 \qquad \text{(C.10d)}$$
$$c_{20} = -p \qquad \text{(C.10e)}$$
$$c_{02} = -p \qquad \text{(C.10f)}$$

from which:

$$x_0 = -\frac{c_{10}}{2\,c_{20}} \qquad \text{(C.11a)}$$
$$y_0 = -\frac{c_{01}}{2\,c_{02}} \qquad \text{(C.11b)}$$
$$k = \exp\left(c_{00} - \frac{c_{10}^2 + c_{01}^2}{4\,c_{20}}\right) \qquad \text{(C.11c)}$$
$$p = -c_{20} = -c_{02}, \ \{c_{20}, c_{02}\} \subset \mathbb{R}^- \qquad \text{(C.11d)}$$
$$\qquad \text{(C.12)}$$

Case II

When $\theta = \pm \pi/4$, from Eqs. (C.4a) to (C.4f):

$$c_{00} = \ln(k) - p(x_0^2 + y_0^2) - 2\,q\,x_0\,y_0 \sin(2\theta) \qquad \text{(C.13a)}$$
$$c_{10} = 2\,p\,x_0 + 2\,q\,y_0 \sin(2\theta) \qquad \text{(C.13b)}$$
$$c_{01} = 2\,p\,y_0 + 2\,q\,x_0 \sin(2\theta) \qquad \text{(C.13c)}$$
$$c_{11} = -2\,q \sin(2\theta) \qquad \text{(C.13d)}$$

$$c_{20} = -p \qquad \text{(C.13e)}$$
$$c_{02} = -p \qquad \text{(C.13f)}$$

from which:

$$x_0 = \frac{2\,c_{10}\,c_{02} - c_{11}\,c_{01}}{\Delta} \qquad \text{(C.14a)}$$
$$y_0 = \frac{2\,c_{20}\,c_{01} - c_{11}\,c_{10}}{\Delta} \qquad \text{(C.14b)}$$
$$k = \exp\left(c_{00} + \frac{c_{10}^2\,c_{02} - c_{10}\,c_{11}\,c_{01} + c_{02}\,c_{01}^2}{\Delta}\right) \qquad \text{(C.14c)}$$
$$p = -c_{20} = -c_{02}, \quad \{c_{20},\,c_{02}\} \subset \mathbb{R}^- \qquad \text{(C.14d)}$$
$$q = -\frac{c_{11}}{2\sin(2\theta)} \qquad \text{(C.14e)}$$

Appendix D

Nonlinear Function Fitting

We consider the problem of fitting a real scalar function f of N real variables x_i and K real parameters β_j to M expected function values using non-linear least-squares optimization with the Levenberg-Marquardt algorithm originally presented by Levenberg (1944) and Marquardt (1963):

$$\{(\overset{k}{x}_1, \ldots, \overset{k}{x}_N, \overset{k}{y}_0)\}, \quad k \in [1, M]_{\mathbb{N}} \tag{D.1}$$

$$y = f(x_i; \beta_j): \mathbb{R}^N \longmapsto \mathbb{R}, \quad i \in [1, N]_{\mathbb{N}}, \quad j \in [1, K]_{\mathbb{N}} \tag{D.2}$$

Levenberg-Marquardt algorithm

With a set of residuals defined as:

$$\overset{k}{R} := \overset{k}{y} - \overset{k}{y}_0 \equiv f(\overset{k}{x}_i; \beta_j) - \overset{k}{y}_0, \quad k \in [1, M]_{\mathbb{N}} \tag{D.3}$$

the problem is to find a set of parameters $\{\beta_j\}, j \in [1, K]_{\mathbb{N}}$, which minimizes:

$$R := \sum_{k=1}^{M} (\overset{k}{R})^2 \tag{D.4}$$

Using the Taylor series expansion over the parameters at a fixed point $(x_i) = (x_1, \ldots, x_N)$ (Bronshtein et al., 2007, p. 417):

$$f(x_i; \beta_j + \delta_j) = f(x_i; \beta_j) + \sum_{k=1}^{K} \frac{\partial f(x_i; \beta_j)}{\partial \beta_k} \delta_k + O[\delta_k^2], \quad \delta_k \in \mathbb{R} \tag{D.5}$$

the residuals and their sum of squares can be approximated by the first order terms of this expansion:

$$\tilde{\overset{k}{R}} := f(\overset{k}{x}_i; \beta_j) + \sum_{m=1}^{K} \left(\frac{\partial f(\overset{k}{x}_i; \beta_j)}{\partial \beta_m} \delta_m \right) - \overset{k}{y}_0 \tag{D.6}$$

$$\tilde{R} := \sum_{k=1}^{M} \left[f(\overset{k}{\vec{x}_i}; \beta_j) + \sum_{m=1}^{K} \left(\frac{\partial f(\overset{k}{\vec{x}_i}; \beta_j)}{\partial \beta_m} \delta_m \right) - \overset{k}{y_0} \right]^2 \tag{D.7}$$

where $(\delta_n) \in \mathbb{R}^K$. If $(\delta_n) \to \mathbf{0}$ then $\tilde{R} \to R$.

Considered as a function of (δ_n), the non-negative function $\tilde{R}$ can be minimized using the standard least-squares method by setting all of its partial derivatives with respect to δ_n equal to zero:

$$\frac{\partial \tilde{R}}{\partial \delta_n} = 0, \ \forall n \in [1, \ K]_\mathbb{N} \tag{D.8}$$

Applying Eq. (D.8) to Eq. (D.7) results in:

$$\sum_{k=1}^{M} \left[\frac{\partial f(\overset{k}{\vec{x}_i}; \beta_j)}{\partial \beta_n} \left(f(\overset{k}{\vec{x}_i}; \beta_j) + \sum_{m=1}^{K} \left(\frac{\partial f(\overset{k}{\vec{x}_i}; \beta_j)}{\partial \beta_m} \delta_m \right) - \overset{k}{y_0} \right) \right] = 0, \ \forall n \in [1, \ K]_\mathbb{N}$$

$$\therefore \quad \sum_{k=1}^{M} \left[\frac{\partial f(\overset{k}{\vec{x}_i}; \beta_j)}{\partial \beta_n} \left(f(\overset{k}{\vec{x}_i}; \beta_j) - \overset{k}{y_0} \right) \right] +$$
$$\sum_{k=1}^{M} \sum_{m=1}^{K} \left(\frac{\partial f(\overset{k}{\vec{x}_i}; \beta_j)}{\partial \beta_n} \cdot \frac{\partial f(\overset{k}{\vec{x}_i}; \beta_j)}{\partial \beta_m} \delta_m \right) = 0, \ \forall n \in [1, \ K]_\mathbb{N} \tag{D.9}$$

The iterative solution of Eq. (D.9) for (δ_m) and updating (β_m) can minimize $\tilde{R}$. For simplicity, the matrix notation is used for the rest of the derivations.

Matrix definitions:

$$\boldsymbol{A} = [A_{mn}]_{M \times K}, \ A_{mn} := \frac{\partial f(\overset{m}{\vec{x}_i}; \beta_j)}{\partial \beta_n}, \ m \in [1, \ M]_\mathbb{N}, \ n \in [1, \ K]_\mathbb{N} \tag{D.10}$$

$$\boldsymbol{B} = [B_{mn}]_{K \times K}, \ B_{mn} := \sum_{k=1}^{M} \left(\frac{\partial f(\overset{k}{\vec{x}_i}; \beta_j)}{\partial \beta_m} \cdot \frac{\partial f(\overset{k}{\vec{x}_i}; \beta_j)}{\partial \beta_n} \right),$$
$$m, n \in [1, \ K]_\mathbb{N} \tag{D.11}$$

$$\boldsymbol{C} = [C_m]_{K \times 1}, \ C_m := \sum_{k=1}^{M} \left(\frac{\partial f(\overset{k}{\vec{x}_i}; \beta_j)}{\partial \beta_m} \left(-f(\overset{k}{\vec{x}_i}; \beta_j) + \overset{k}{y_0} \right) \right),$$
$$m \in [1, \ K]_\mathbb{N} \tag{D.12}$$

Appendix D. Nonlinear Function Fitting

$$\boldsymbol{D} = [D_m]_{K \times 1}, \quad D_m := \delta_m, \quad m \in [1, \, K]_{\mathbb{N}} \tag{D.13}$$

$$\boldsymbol{E} = [E_m]_{M \times 1}, \quad E_m := -f(\overset{m}{x_i};\, \beta_j) + \overset{m}{y}_0, \quad m \in [1, \, M]_{\mathbb{N}} \tag{D.14}$$

From Eqs. (D.10) and (D.11):

$$\boldsymbol{B} = \boldsymbol{A}^T \cdot \boldsymbol{A} \tag{D.15}$$

and from Eqs. (D.14) and (D.15):

$$\boldsymbol{C} = (\boldsymbol{E}^T \cdot \boldsymbol{A})^T = \boldsymbol{A}^T \cdot \boldsymbol{E} \tag{D.16}$$

where the superscript T denotes the matrix transposition. With these definitions, Eqs. (D.7) and (D.9) in the matrix notation are:

$$\tilde{R} = (\boldsymbol{A} \cdot \boldsymbol{D} - \boldsymbol{E})^T \cdot (\boldsymbol{A} \cdot \boldsymbol{D} - \boldsymbol{E}) \tag{D.17}$$

$$\boldsymbol{B} \cdot \boldsymbol{D} = \boldsymbol{C} \tag{D.18}$$

Theorem D.1 (Marquardt, 1963, p. 434) Let $\lambda \geq 0$ be arbitrary and let

$$\boldsymbol{D}_0 = [D_{0,m}]_{K \times 1}, \quad D_{0,m} := \delta_{0,m}, \quad m \in [1, \, K]_{\mathbb{N}} \tag{D.19}$$

satisfy the equation:

$$(\boldsymbol{B} + \lambda \boldsymbol{I}) \cdot \boldsymbol{D}_0 = \boldsymbol{C} \tag{D.20}$$

Then $\boldsymbol{D}_0$ minimizes $\tilde{R}$ on a sphere in $\mathbb{R}^K$, whose radius $\sqrt{\boldsymbol{D}^T \cdot \boldsymbol{D}}$ satisfies:

$$\boldsymbol{D}^T \cdot \boldsymbol{D} = \boldsymbol{D}_0^T \cdot \boldsymbol{D}_0 \tag{D.21}$$

Proof: In order to find a vector $\boldsymbol{D}$, which optimizes Eq. (D.17) under the constraint (D.21), the Lagrange method requires:

$$\frac{\partial u}{\partial D_n} = \frac{\partial u}{\partial \delta_n} = 0, \quad \forall n \in [1, K]_{\mathbb{N}} \tag{D.22}$$

where

$$u(\delta_i, \, \lambda) = u(\boldsymbol{D}, \, \lambda) := \tilde{R} + \lambda(\boldsymbol{D}^T \cdot \boldsymbol{D} - \boldsymbol{D}_0^T \cdot \boldsymbol{D}_0) \tag{D.23}$$

and $\lambda \in \mathbb{R}$ is the Lagrange multiplier. Equation (D.22) together with Eq. (D.21) are the necessary conditions for $\tilde{R}$ to become stationary.

From Eqs. (D.7) and (D.13):

$$u = \sum_{k=1}^{M} \left[f(\overset{k}{x}_i; \beta_j) + \sum_{m=1}^{K} \left(\frac{\partial f(\overset{k}{x}_i; \beta_j)}{\partial \beta_m} \delta_m \right) - \overset{k}{y}_0 \right]^2 + \lambda \cdot \sum_{m=1}^{K} \left(\delta_m^2 - \delta_{0,m}^2 \right) \qquad (D.24)$$

From Eqs. (D.22) and (D.24):

$$\sum_{k=1}^{M} \left[\frac{\partial f(\overset{k}{x}_i; \beta_j)}{\partial \beta_n} \left(f(\overset{k}{x}_i; \beta_j) + \sum_{m=1}^{K} \left(\frac{\partial f(\overset{k}{x}_i; \beta_j)}{\partial \beta_m} \delta_m \right) - \overset{k}{y}_0 \right) \right]$$

$$+ \lambda \delta_n = 0, \quad \forall n \in [1, K]_{\mathbb{N}}$$

$$\therefore \sum_{k=1}^{M} \left[\frac{\partial f(\overset{k}{x}_i; \beta_j)}{\partial \beta_n} \left(f(\overset{k}{x}_i; \beta_j) - \overset{k}{y}_0 \right) \right]$$

$$+ \sum_{k=1}^{M} \sum_{m=1}^{K} \left(\frac{\partial f(\overset{k}{x}_i; \beta_j)}{\partial \beta_n} \cdot \frac{\partial f(\overset{k}{x}_i; \beta_j)}{\partial \beta_m} \delta_m \right) + \lambda \delta_n = 0,$$

$$\forall n \in [1, K]_{\mathbb{N}} \qquad (D.25)$$

Equation (D.25) in matrix form is:

$$(\boldsymbol{B} + \lambda \boldsymbol{I}) \cdot \boldsymbol{D}_0 = \boldsymbol{C} \qquad (D.26)$$

in which $\boldsymbol{D}_0$ is the value of $\boldsymbol{D}$ for which Eq. (D.26) holds.

Expanding Eq. (D.17) by its components gives:

$$\tilde{R} = \sum_{i=1}^{M} \left(\sum_{j=1}^{K} (A_{ij} D_j) - E_i \right)^2 \qquad (D.27)$$

For $\boldsymbol{D} = \boldsymbol{0}$, $\tilde{R} = \boldsymbol{E}^T \cdot \boldsymbol{E} \geq 0$. Starting from $\boldsymbol{0}$, $\boldsymbol{D}$ can be arbitrary chosen to reduce $\tilde{R}$. It can also be chosen so that $\tilde{R} \to +\infty$. Since $\tilde{R}$ is non-negative and continuous, for a continuous variation of $\boldsymbol{D}$, $\boldsymbol{D}_0$ in Eq. (D.26) minimizes $\tilde{R}$. □

Scaling of variables

In the following, scaled variables and parameters are denoted by "$*$" as superscript.

From Eq. (D.26):

$$\sum_{n=1}^{K} \left[\left(\frac{B_{mn}}{\sqrt{B_{mm}}\sqrt{B_{nn}}} + \frac{\lambda I_{mn}}{\sqrt{B_{mm}}\sqrt{B_{nn}}} \right) D_{0,n} \sqrt{B_{nn}} \right] = \frac{C_m}{\sqrt{B_{mm}}},$$
$$\forall m \in [1,\ K]_{\mathbb{N}} \tag{D.28}$$

$$\therefore \quad \sum_{n=1}^{K} \left[(B_{mn}^* + \lambda I_{mn}^*) D_{0,n}^* \right] = C_m^*, \quad \forall m \in [1,\ K]_{\mathbb{N}} \tag{D.29}$$

or $\quad (\boldsymbol{B}^* + \lambda \boldsymbol{I}^*) \cdot \boldsymbol{D}_0^* = \boldsymbol{C}^* \tag{D.30}$

where:

$$\boldsymbol{B}^* = [B_{mn}^*]_{K \times K} := \left[\frac{B_{mn}}{\sqrt{B_{mm}}\sqrt{B_{nn}}} \right]_{K \times K} \tag{D.31}$$

$$\boldsymbol{I}^* = [I_{mn}^*]_{K \times K} := \left[\frac{I_{mn}}{\sqrt{B_{mm}}\sqrt{B_{nn}}} \right]_{K \times K} \tag{D.32}$$

$$\boldsymbol{C}^* = [C_m^*]_{K \times 1} := \left[\frac{C_m}{\sqrt{B_{mm}}} \right]_{K \times 1} \tag{D.33}$$

$$\boldsymbol{D}^* = [D_n^*]_{K \times 1} := \left[D_n \sqrt{B_{nn}} \right]_{K \times 1} \quad \Leftrightarrow \quad \boldsymbol{D} = \left[\frac{D_n^*}{\sqrt{B_{nn}}} \right]_{K \times 1} \tag{D.34}$$

Appendix E

Discretization

For a rectangular structured grid in the two-dimensional Cartesian coordinates $\xi_1 \xi_2$ with constant grid spacings δ_1 and δ_2 along ξ_1 and ξ_2 axes, Fig. E.1, the discretized derivatives of a real function $u(\xi_1, \xi_2)$ are as follows (Tannehill et al., 1997):

$$u_{,1}|_{i,j} = \frac{u_{i+1,j} - u_{i-1,j}}{2\delta_1} + O(\delta_1^2) \tag{E.1}$$

$$u_{,1}|_{i,j} = \frac{-3\, u_{i,j} + 4\, u_{i+1,j} - u_{i+2,j}}{2\delta_1} + O(\delta_1^2) \tag{E.2}$$

$$u_{,1}|_{i,j} = \frac{3\, u_{i,j} - 4\, u_{i-1,j} + u_{i-2,j}}{2\delta_1} + O(\delta_1^2) \tag{E.3}$$

$$u_{,11}|_{i,j} = \frac{u_{i+1,j} - 2\, u_{i,j} + u_{i-1,j}}{\delta_1^2} + O(\delta_1^2) \tag{E.4}$$

$$u_{,11}|_{i,j} = \frac{2\, u_{i,j} - 5\, u_{i+1,j} + 4\, u_{i+2,j} - u_{i+3,j}}{\delta_1^2} + O(\delta_1^2) \tag{E.5}$$

$$u_{,11}|_{i,j} = \frac{2\, u_{i,j} - 5\, u_{i-1,j} + 4\, u_{i-2,j} - u_{i-3,j}}{\delta_1^2} + O(\delta_1^2) \tag{E.6}$$

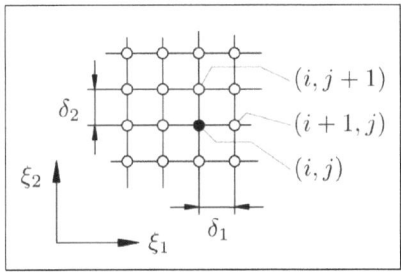

Fig. E.1: The definitions of the discretization on a rectangular grid

$$u_{,12}\big|_{i,j} = \frac{u_{i+1,j+1} - u_{i+1,j-1} - u_{i-1,j+1} + u_{i-1,j-1}}{4\,\delta_1\,\delta_2} + O(\delta_1^2,\,\delta_2^2) \tag{E.7}$$

$$u_{,12}\big|_{i,j} = \frac{u_{i+1,j+1} - u_{i+1,j-1} - u_{i,j+1} + u_{i,j-1}}{2\,\delta_1\,\delta_2} + O(\delta_1,\,\delta_2^2) \tag{E.8}$$

$$u_{,12}\big|_{i,j} = \frac{u_{i,j+1} - u_{i,j-1} - u_{i-1,j+1} + u_{i-1,j-1}}{2\,\delta_1\,\delta_2} + O(\delta_1,\,\delta_2^2) \tag{E.9}$$

$$u_{,12}\big|_{i,j} = \frac{u_{i+1,j+1} - u_{i+1,j} - u_{i-1,j+1} + u_{i-1,j}}{2\,\delta_1\,\delta_2} + O(\delta_1^2,\,\delta_2) \tag{E.10}$$

$$u_{,12}\big|_{i,j} = \frac{u_{i+1,j} - u_{i+1,j-1} - u_{i-1,j} + u_{i-1,j-1}}{2\,\delta_1\,\delta_2} + O(\delta_1^2,\,\delta_2) \tag{E.11}$$

$$u_{,12}\big|_{i,j} = \frac{u_{i+1,j} - u_{i+1,j-1} - u_{i,j} + u_{i,j-1}}{\delta_1\,\delta_2} + O(\delta_1,\,\delta_2) \tag{E.12}$$

$$u_{,12}\big|_{i,j} = \frac{u_{i,j+1} - u_{i,j} - u_{i-1,j+1} + u_{i-1,j}}{\delta_1\,\delta_2} + O(\delta_1,\,\delta_2) \tag{E.13}$$

$$u_{,12}\big|_{i,j} = \frac{u_{i,j} - u_{i,j-1} - u_{i-1,j} + u_{i-1,j-1}}{\delta_1\,\delta_2} + O(\delta_1,\,\delta_2) \tag{E.14}$$

$$u_{,12}\big|_{i,j} = \frac{u_{i+1,j+1} - u_{i+1,j} - u_{i,j+1} + u_{i,j}}{\delta_1\,\delta_2} + O(\delta_1,\,\delta_2) \tag{E.15}$$

where a comma in the subscript denotes partial derivatives, for example:

$$u_{,12} := \frac{\partial u(\xi_1,\xi_2)}{\partial \xi_1 \partial \xi_2} \tag{E.16}$$

The discretized forms of $u_{,2}$ and $u_{,22}$ are derived from $u_{,1}$ and $u_{,11}$ in the above equations by replacing δ_1 with δ_2 and interchanging the increments of the indices i and j.

The order of the discretization equations is two at the internal nodes and two or mixed one and two at the boundary nodes.

Appendix F

Calibration Correlations of 5-hole Pressure Probes

The correlation functions for the calibration of the 5-hole pressure probes, Eqs. (4.1) to (4.5), in the *null-reading* mode of measurements are as follows:

The yaw angle, α, and its offset, α_0, are approximately independent from the Mach number:

Probe No. 3190: $M \in [0.1939012, 0.6836150]$

$$\alpha(C_{YA}, M) = -C_{YA}/0.04293 \qquad \text{(F.1)}$$
$$\alpha_0(\gamma, M) = 1.02248 \times 10^{-4} \gamma^2 + 0.0171408\,\gamma - 1.11850 \qquad \text{(F.2)}$$

Probe No. 3192: $M \in [0.1856182, 0.69668]$

$$\alpha(C_{YA}, M) = -C_{YA}/0.04293 \qquad \text{(F.3)}$$
$$\alpha_0(\gamma, M) = -9 \times 10^{-7} \gamma^4 + 10^{-6} \gamma^3 + 8 \times 10^{-5} \gamma^2$$
$$- 0.0044\,\gamma + 4.2298 \qquad \text{(F.4)}$$

Probe No. 3194: $M \in [0.185459704, 0.705275306]$

$$\alpha(C_{YA}, M) = -24.38\,C_{YA} \qquad \text{(F.5)}$$
$$\alpha_0(\gamma, M) = 0.0003\,\gamma^2 + 0.0079\,\gamma - 1.8 \qquad \text{(F.6)}$$

The equations of the offset angle α_0 are the correlations of the differences between the geometric and the aerodynamic yaw angles. It is subtracted from the measured yaw angle to give the geometric yaw angle.

The correlation functions of the pitch angle and the pressure coefficients are as follows:

$$\gamma(C_{PA}, M) = \sum_{i=0}^{6} a_i(M)\, C_{PA}^i \quad \text{(Table F.1)} \tag{F.7}$$

$$C_{DP}(\gamma, M) = \sum_{i=0}^{5} b_i(M)\, \gamma^i \quad \text{(Table F.2)} \tag{F.8}$$

$$C_{SP}(\gamma, M) = \sum_{i=0}^{4} c_i(M)\, \gamma^i \quad \text{(Table F.3)} \tag{F.9}$$

$$C_{TP}(\gamma, M) = \sum_{i=0}^{5} d_i(M)\, \gamma^i \quad \text{(Table F.4)} \tag{F.10}$$

All angles in the above equations are in degrees. The calibration range of the pitch angle for the pressure probes 3190 and 3192 is $\gamma \in [-40°, +40°]$ and for the probe 3094 is $\gamma \in [-49°, +49°]$.

The coefficients of the calibration correlations of the pressure probes 3190, 3192, and 3194 are given in the following tables.

Appendix F. Calibration Correlations of 5-hole Pressure Probes 133

Table F.1: The coefficients of the pitch angle correlation, Eq. (F.7)

M	$a_6(M)$	$a_5(M)$	$a_4(M)$	$a_3(M)$	$a_2(M)$	$a_1(M)$	$a_0(M)$
Probe: 3190							
0.1939012	0.0	-40.35255	+9.791615	+69.85431	-8.881632	-74.26935	-1.037767
0.2971538	0.0	-39.01608	+12.41779	+67.52198	-10.24983	-73.40699	-1.286507
0.4529735	0.0	-41.57831	+11.20979	+68.60669	-8.873537	-71.79849	-1.686656
0.6006941	0.0	-34.65609	+11.32417	+59.37358	-8.677521	-65.88210	-1.899256
0.6836150	0.0	-41.78831	+18.45139	+58.08616	-10.39595	-61.54189	-1.926609
Probe: 3192							
0.1856182	+70.277	-22.774	-101.16	+40.348	+48.585	-72.519	-12.629
0.3017698	+111.48	+3.0102	-112.98	+44.368	+48.343	-76.833	-11.854
0.431311	+43.369	-55.596	-75.712	+70.967	+40.727	-76.721	-10.394
0.605941	+61.607	-54.775	-84.308	+71.708	+39.365	-73.797	-9.2559
0.69668	+77.787	-103.84	-68.734	+87.404	+36.075	-73.107	-9.3391
Probe: 3194							
0.185459704	0.0	0.0	0.0	0.0	7.40747	-69.1117	+0.8967410
0.299448952	0.0	0.0	0.0	0.0	3.97168	-69.3155	+1.0198900
0.447947891	0.0	0.0	0.0	0.0	1.41475	-67.5631	+0.5916590
0.611408367	0.0	0.0	0.0	0.0	-1.71097	-65.2393	-0.444
0.705275306	0.0	0.0	0.0	0.0	7.93469	-66.5562	-1.2746200

Table F.2: The coefficients of the dynamic pressure correlation, Eq. (F.8)

M	$b_5(M)$	$b_4(M)$	$b_3(M)$	$b_2(M)$	$b_1(M)$	$b_0(M)$
Probe: 3190						
0.1939012	0.0	-1.271796E-09	+7.689452E-08	-3.149368E-07	-1.416386E-04	2.172675E-02
0.2971538	0.0	-1.910896E-09	+1.695330E-07	-2.106573E-06	-2.920607E-04	4.770339E-02
0.4529735	0.0	-6.175179E-09	+2.912469E-07	-2.614237E-07	-5.046550E-04	1.027185E-01
0.6006941	0.0	-6.094900E-09	+4.806842E-07	-7.471407E-06	-8.280554E-04	1.641070E-01
0.6836150	0.0	-8.636687E-09	+5.737960E-07	-4.792790E-06	-1.025619E-03	1.950618E-01
Probe: 3192						
0.1856182	-2E-11	-6E-10	+8E-08	-3E-07	-9E-05	+0.0208
0.3017698	-9E-13	-2E-10	+1E-07	-3E-06	-0.0002	+0.0499
0.431311	-9E-12	-2E-09	+2E-07	-4E-06	-0.0003	+0.1071
0.605941	-9E-12	-9E-10	+4E-07	-1E-05	-0.0005	+0.1754
0.69668	1E-10	-3E-09	+2E-07	-7E-06	-0.0006	+0.213
Probe: 3194						
0.185459704	0.0	-4.7374E-10	+3.1900E-08	+3.2296E-07	-4.6355E-05	+2.21E-02
0.299448952	0.0	+1.23787E-10	+3.93440E-08	-2.21834E-06	-8.32112E-05	+5.54E-02
0.447947891	0.0	-2.6527E-09	+2.8801E-08	+3.9616E-07	-1.0688E-04	+1.20E-01
0.611408367	0.0	+9.0989E-10	+2.0502E-07	-5.4623E-06	-4.0243E-04	+2.02E-01
0.705275306	0.0	-4.4527E-09	+1.0194E-07	+1.5034E-06	-3.2614E-04	+2.55E-01

Appendix F. Calibration Correlations of 5-hole Pressure Probes   135

Table F.3: The coefficients of the static pressure correlation, Eq. (F.9)

M	$c_4(M)$	$c_3(M)$	$c_2(M)$	$c_1(M)$	$c_0(M)$
Probe: 3190					
0.1939012	+8.512461E-08	-4.305837E-06	+1.585613E-06	+8.514497E-03	+0.2200743
0.2971538	+5.027508E-08	-4.601377E-06	+7.797571E-05	+8.468228E-03	+0.2640518
0.4529735	+7.823675E-08	-3.665374E-06	+2.121983E-05	+6.519854E-03	+0.2833294
0.6006941	+4.670087E-08	-4.090756E-06	+8.161113E-05	+7.225924E-03	+0.3339905
0.6836150	+5.437153E-08	-4.044247E-06	+5.671979E-05	+7.332332E-03	+0.3895203
Probe: 3192					
0.1856182	0.0	-3E-06	+6E-05	+0.0048	+0.169
0.3017698	0.0	-2E-06	+8E-05	+0.0033	+0.1676
0.431311	0.0	-3E-06	+8E-05	+0.005	+0.2404
0.605941	0.0	-3E-06	+1E-04	+0.0043	+0.264
0.69668	0.0	-3E-06	+8E-05	+0.0044	+0.2973
Probe: 3194					
0.185459704	+2.48824E-08	-1.27591E-06	-3.18356E-05	+2.20509E-03	+9.07E-02
0.299448952	-8.5824E-09	-4.9689E-07	+3.7846E-05	+1.4893E-03	+9.83E-02
0.447947891	+1.94926E-08	+8.29384E-08	-1.10352E-05	+6.92042E-04	+7.40E-02
0.611408367	-1.5869E-08	-1.0106E-06	+3.4229E-05	+2.1588E-03	+1.10E-01
0.705275306	+1.1794E-08	-2.2791E-07	-3.1123E-06	+1.2092E-03	+1.09E-01

Table F.4: The coefficients of the total pressure correlation, Eq. (F.10)

M	$d_5(M)$	$d_4(M)$	$d_3(M)$	$d_2(M)$	$d_1(M)$	$d_0(M)$
Probe: 3190						
0.1939012	0.0	0.0	-3.097339E-07	-1.640291E-05	+4.979787E-04	+2.370160E-02
0.2971538	0.0	-7.567600E-09	-1.173732E-07	+1.056577E-05	+3.553459E-04	+1.603981E-02
0.4529735	0.0	-1.046248E-08	-1.112873E-07	+1.667853E-05	+4.329132E-05	+3.482588E-03
0.6006941	0.0	-1.350092E-08	-2.641646E-07	+1.297179E-05	+2.295971E-04	+1.748392E-02
0.6836150	0.0	-1.001014E-08	-8.478926E-08	-4.548147E-06	-2.841884E-04	+1.647617E-02
Probe: 3192						
0.1856182	-6E-10	-4E-09	+9E-07	-3E-06	+2E-05	+0.0335
0.3017698	+2E-10	-1E-08	-6E-07	+1E-05	+0.0003	+0.0109
0.431311	+5E-11	-8E-09	-4E-07	+7E-06	+0.0003	+0.0588
0.605941	-2E-10	-2E-08	+5E-08	+2E-05	+5E-05	+0.0134
0.69668	-3E-10	-1E-08	+8E-08	+2E-05	-0.0002	-0.0006
Probe: 3194						
0.185459704	0.0	+1.4178E-09	+2.5959E-07	-1.6651E-05	-3.5340E-05	+1.83E-02
0.299448952	0.0	-8.23839E-09	+2.43118E-07	-7.18602E-06	-1.79302E-04	+1.05E-02
0.447947891	0.0	-5.73862E-09	+3.6207E-07	-8.5615E-06	-2.9490E-04	+4.76E-03
0.611408367	0.0	-1.3806E-08	+9.3093E-08	+3.0019E-06	-7.2720E-05	+9.81E-03
0.705275306	0.0	-9.4989E-09	+2.7263E-07	+2.3972E-06	-2.8057E-04	+4.61E-03

Appendix G

Calibration Charts

This appendix presents the calibration charts of the ambient absolute pressure transmitter and the pressure transmitters used in the 5-hole probe measurements. The channel assignments to the pressure probes and the zero-drift of the pressure channels are given in the following tables. The nomenclature of the probe pressures is given in Fig. 4.8a.

Table G.1: Channel assignments of the pressure probes

Probe Pressure	Channel Number
Probe No. 3190: Plane 1	
p_1	301
p_2	302
$p_2 - p_3$	303
$p_4 - p_5$	305
Probe No. 3194: Plane 2	
p_1	306
p_2	307
$p_2 - p_3$	308
$p_4 - p_5$	309
Probe No. 3192: Plane 3	
p_1	310
p_2	311
$p_2 - p_3$	312
$p_4 - p_5$	313

Table G.2: Maximum zero-drifts of the pressure channels during 24 hours

Channel No.	Channel Value [VDC]	Channel No.	Channel Value [VDC]
201	±0.002	301	±0.0015
202	±0.004	302	±0.002
203	±0.001	303	±0.001
204	±0.001	305	±0.0045
205	±0.002	306	±0.003
206	±0.003	307	±0.0005
207	±0.002	308	±0.001
211	±0.001	309	±0.0015
212	±0.001	310	±0.0025
213	±0.003	311	±0.0006
		312	±0.0005
		313	±0.001

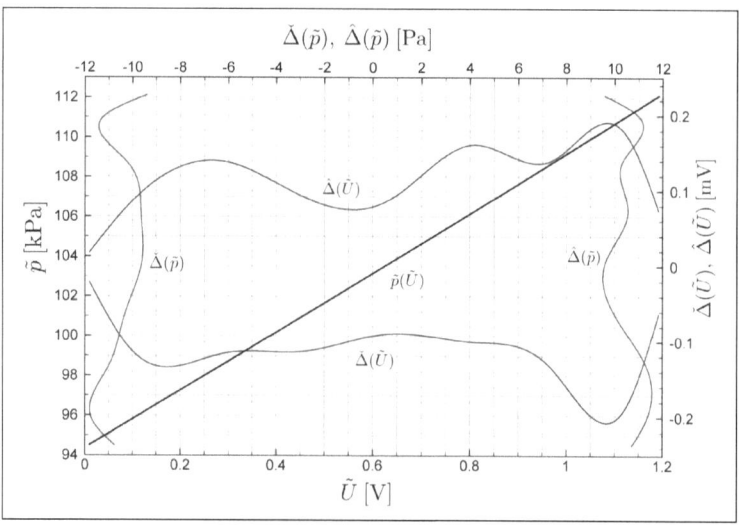

Fig. G.1: The Barton-cell absolute pressure transmitter for the measurement of the ambient pressure

Appendix G. Calibration Charts

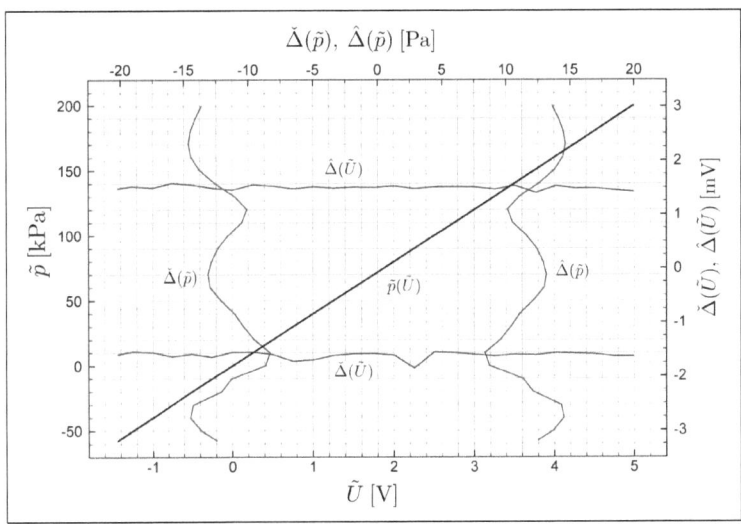

Fig. G.2: Pressure channel 301, corresponding to the pressure p_1 of the probe in plane 1

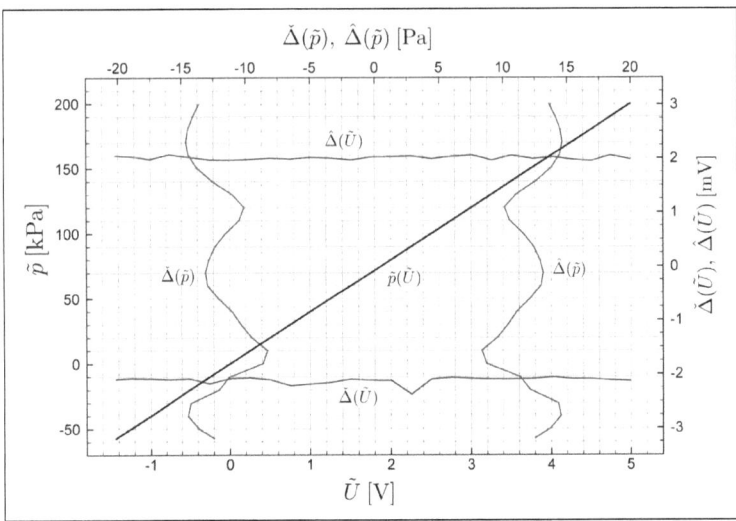

Fig. G.3: Pressure channel 302, corresponding to the pressure p_2 of the probe in plane 1

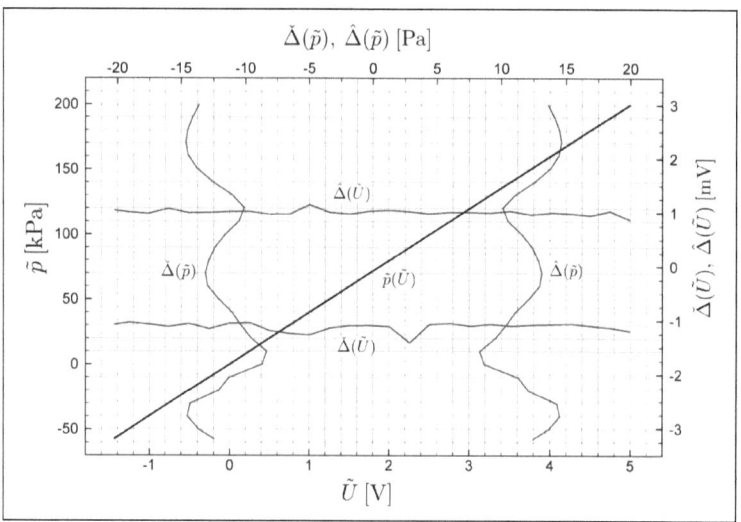

Fig. G.4: Pressure channel 303, corresponding to the pressure difference $p_2 - p_3$ of the probe in plane 1

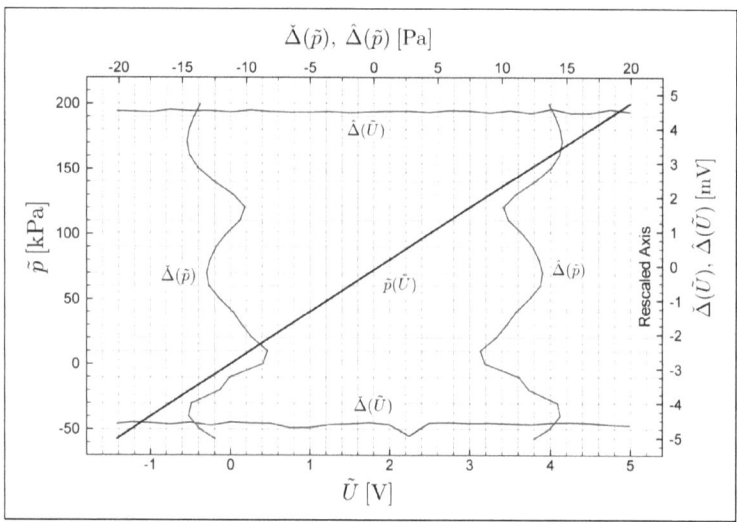

Fig. G.5: Pressure channel 305, corresponding to the pressure difference $p_4 - p_5$ of the probe in plane 1

Appendix G. Calibration Charts

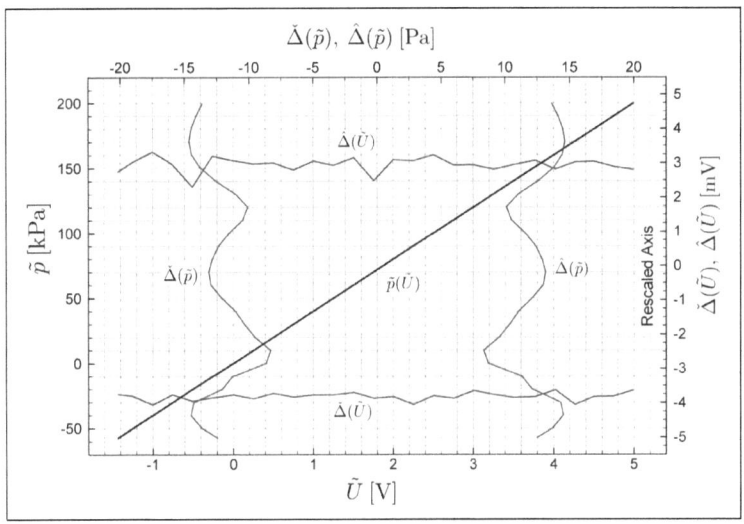

Fig. G.6: Pressure channel 306, corresponding to the pressure p_1 of the probe in plane 2

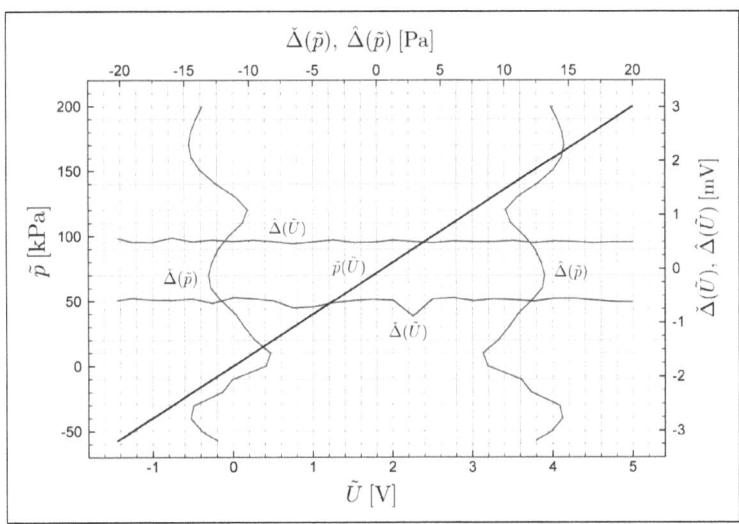

Fig. G.7: Pressure channel 307, corresponding to the pressure p_2 of the probe in plane 2

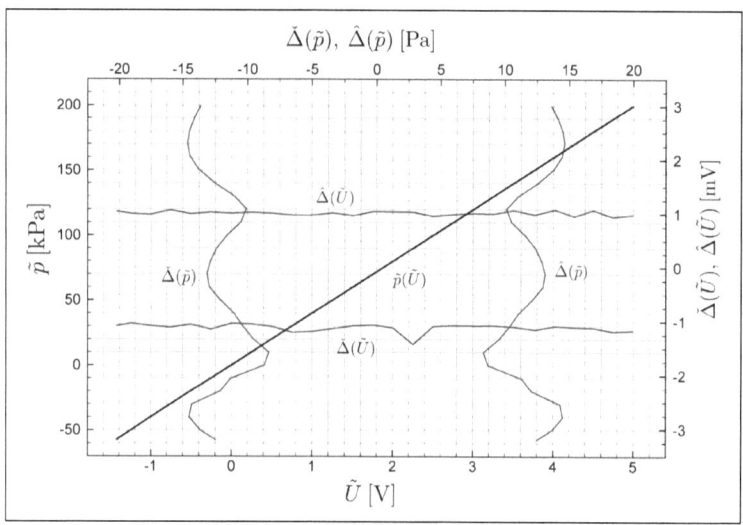

Fig. G.8: Pressure channel 308, corresponding to the pressure difference $p_2 - p_3$ of the probe in plane 2

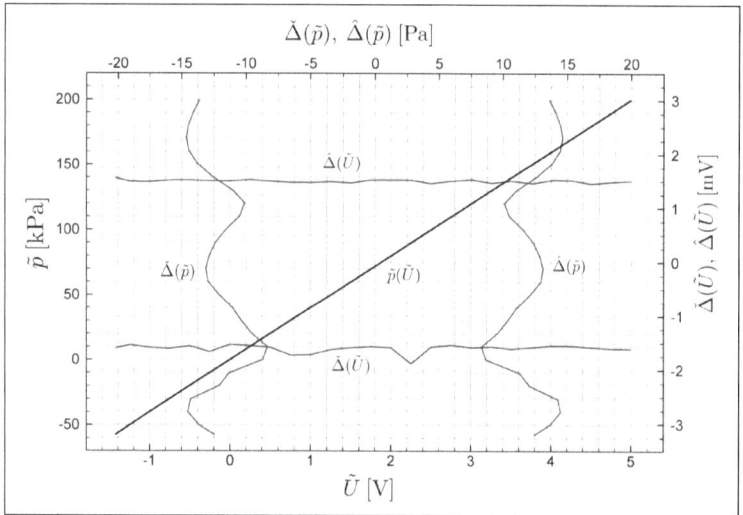

Fig. G.9: Pressure channel 309, corresponding to the pressure difference $p_4 - p_5$ of the probe in plane 2

Appendix G. Calibration Charts

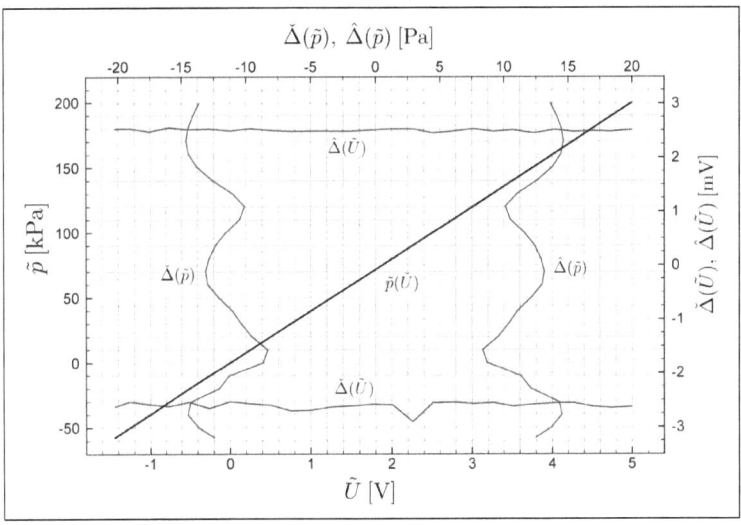

Fig. G.10: Pressure channel 310, corresponding to the pressure p_1 of the probe in plane 3

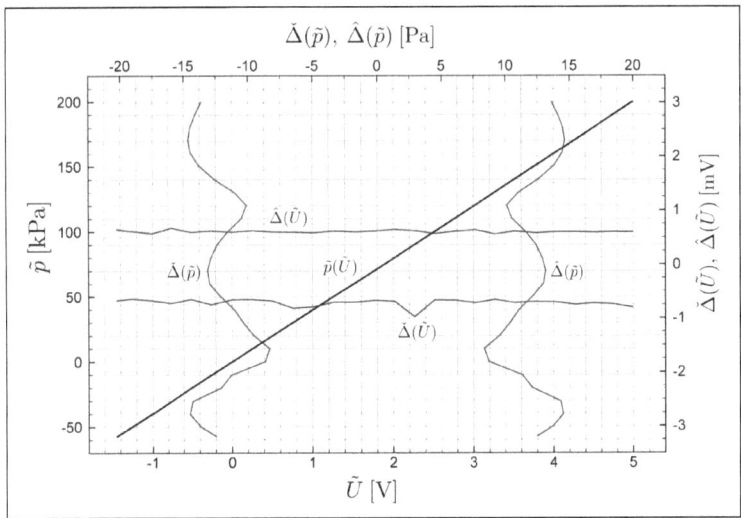

Fig. G.11: Pressure channel 311, corresponding to the pressure p_2 of the probe in plane 3

144　　　　　　　　　　　　　　　　　　　　　　　　Appendix G. Calibration Charts

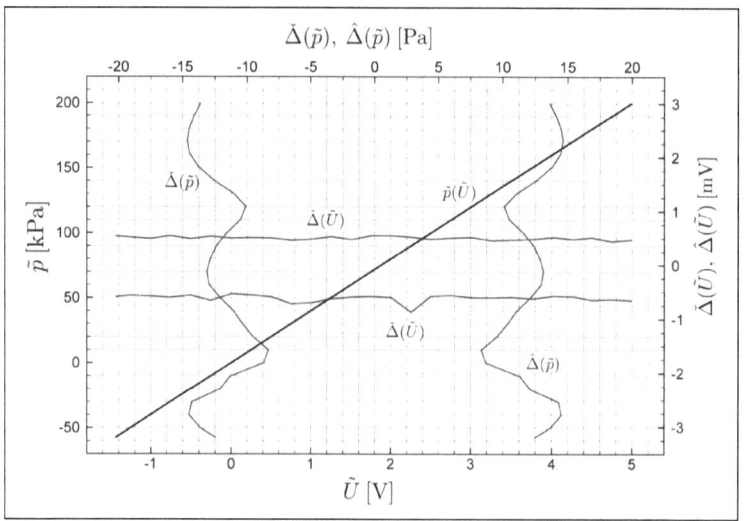

Fig. G.12: Pressure channel 312, corresponding to the pressure difference $p_2 - p_3$ of the probe in plane 3

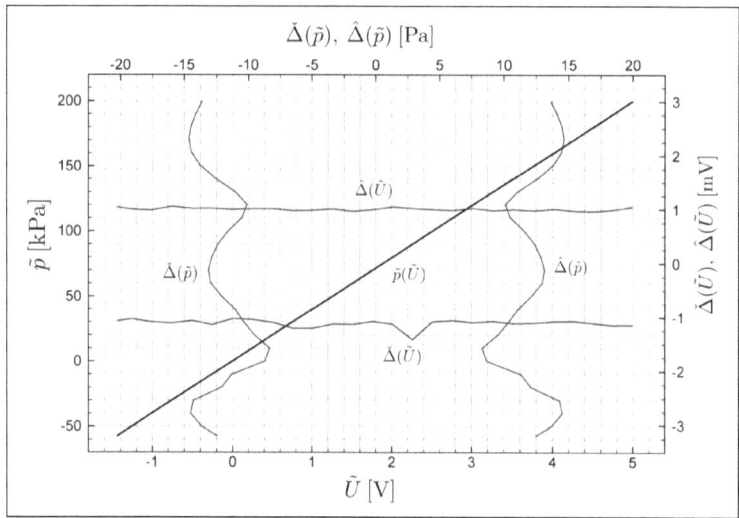

Fig. G.13: Pressure channel 313, corresponding to the pressure difference $p_4 - p_5$ of the probe in plane 3

I want morebooks!

Buy your books fast and straightforward online - at one of world's fastest growing online book stores! Environmentally sound due to Print-on-Demand technologies.

Buy your books online at
www.morebooks.shop

Kaufen Sie Ihre Bücher schnell und unkompliziert online – auf einer der am schnellsten wachsenden Buchhandelsplattformen weltweit! Dank Print-On-Demand umwelt- und ressourcenschonend produziert.

Bücher schneller online kaufen
www.morebooks.shop

KS OmniScriptum Publishing
Brivibas gatve 197
LV-1039 Riga, Latvia
Telefax: +371 686 204 55

info@omniscriptum.com
www.omniscriptum.com

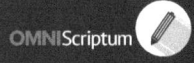

Printed by Books on Demand GmbH, Norderstedt / Germany